THE SACRED MUSHROOM

"In *The Sacred Mushroom*, Andrija Puharich takes us on an extraordinary journey into a realm where ancient whispers intertwine with the pulse of modern exploration. The *Amanita muscaria*—mysterious and powerful, a gateway to the unseen—reveals its long-veiled truths through his words, weaving a path through luminous corridors of consciousness and awakening. A beacon for seekers, a key for the curious, this book dissolves the walls of the ordinary, unveiling the infinite mystery that lies beyond. A must-read for those drawn to the sacred dance of mysticism, science, and the eternal unknown."

BABA MASHA, M.D., AUTHOR OF
MICRODOSING WITH AMANITA MUSCARIA

THE SACRED MUSHROOM

KEY TO THE DOOR OF ETERNITY

ANDRIJA PUHARICH

WITH AN INTRODUCTION BY
P. D. NEWMAN

Park Street Press
Rochester, Vermont

Park Street Press
One Park Street
Rochester, Vermont 05767
www.ParkStPress.com

Park Street Press is a division of Inner Traditions International

Copyright © 1959, 2025 by the Estate of Andrija Puharich
Introduction © 2025 by P. D. Newman

Originally published in 1959 by Doubleday & Company
Revised edition published in 2025 by Park Street Press

All rights reserved. No part of this book may be reproduced or utilized in any form or by any means, electronic or mechanical, including photocopying, recording, or any information storage and retrieval system, without permission in writing from the publisher. No part of this book may be used or reproduced to train artificial intelligence technologies or systems.

Cataloging-in-Publication Data for this title is available from the Library of Congress

ISBN 978-1-64411-687-6 (print)
ISBN 978-1-64411-688-3 (ebook)

Printed and bound in the United States by Lake Book Manufacturing, LLC

10 9 8 7 6 5 4 3 2 1

Scan the QR code and save 25% at InnerTraditions.com. Browse over 2,000 titles on spirituality, the occult, ancient mysteries, new science, holistic health, and natural medicine.

Contents

Introduction to the New Edition   1
by *P. D. Newman*

THE ORIGINAL TEXT
THE SACRED MUSHROOM
THE KEY TO THE DOOR OF ETERNITY   7

Index   223

About the Author   231

Introduction to the New Edition

Henry Karel Puharić, better known as Andrija (Croatian for "Andrew") Puharich, is perhaps best remembered as the man who introduced Israeli psychic Uri Geller to the world. But long before Puharich was documenting Geller's knack for E.T. contact, mind reading, and spoon bending, Puharich was working as a medical officer for the United States government. It was in 1954, while stationed at the Army Chemical Center in Edgewood, Maryland, when an ancient, visionary substance presaging future things to come landed luckily and unexpectedly in his lap. On June 17, a strange event involving a Dutch sculptor and telepath named Harry Stone was casually brought to Puharich's attention. The previous night (June 16), American heiress and member of the Astor family Ava Alice Pleydell-Bouverie was hosting a dinner party at her home on East 61st Street in New York City, at which Harry Stone was present. Thinking that as an artist, Stone would be interested in the piece, the quinquagenarian socialite handed the young sculptor a gold pendant—jewelry that had allegedly belonged to Tiye, the Great Royal Wife of Amenhotep III, mother of Akhenaten, and grandmother of Tutankhamun. However, rather than offering a comment on the artifact, as soon as Bouverie had handed him the piece, Stone immediately

2 INTRODUCTION TO THE NEW EDITION

"trembled all over, got a crazy staring look in his eye, staggered around the room a bit, and fell into a chair," apparently in a state of somnambulistic trance—at which point he requested a pencil and some paper. In addition to a number of what appeared to be ancient Egyptian hieroglyphs, Stone proceeded to draw what can only be described as an anatomically correct fruiting body of the infamous, intoxicating mycorrhizal mushroom *Amanita muscaria*, aka fly agaric. As he sketched, he spoke:

> *TEHUTI AKH. NESI NESU KHUTA NEFERT KUFA ANKH KHUT. PTAH KHUFU. PTAH KATU. EN KATU.* [...] *There is a cream to take people out of themselves when they couldn't bear their pain. The same way (be very careful) as a plant I see here. It grows here and works the same way.* [...] *The spots are white. The plant is yellow. Take the skins off its neck, and the white spots too, they will work the same way. The ointment is to be rubbed in the skull on the joints for trance. If dose is too much you can get in trance and may never wake up. You are so changed in body that you can't get back. Be very careful. Start out with very little and work up. Find what you can stand.*

While the intoxicating properties of *A. muscaria* have been known since at least 1730 with the initial publication of Johan von Strahlenberg's book *An Historical and Geographical Description of the North and Eastern Parts of Europe and Asia, Particularly of Russia, Siberia, and Tartary*—a text concerning a region where the concept of shamanic flight is commonplace—Stone's communication marks the first time that the motif of soul flight was ever directly associated with the fly agaric (or any other) mushroom. Moreover, the connection signals the beginnings of a long trend in historical, anthropological, mycological, and botanical research that seeks to explain magical and mystical tropes in terms of the use of psychedelic drugs. Perhaps more significantly, unlike later arguments for the use of such substances in magico-religious settings that relied heavily upon solid evidence, sound deduction, and logical reasoning, Stone's transmission instead depended solely upon an apparent act of entranced channeling—and that from an alleged four-thousand-six-hundred-year-old high-born Egyptian. To say that Puharich had stumbled upon an interesting area of research would be putting it lightly. His interest was piqued.

Turning his attention to the hieroglyphic symbols, Puharich enlisted a medical associate with a degree in Egyptology in an attempt to determine whether or not Stone's speech and sketches were legitimate. Not only was his colleague able to successfully translate the message, but he noted that many of the symbols employed represented rather archaic glyphs that, having been dropped shortly after the time of the early dynasties, were rarely used in Egyptian writing. "As far as I know," the translator remarked, "there are only a few scholars in the United States who are expert in this particular period of Egyptian history, so-called 'Old Kingdom.'" Among the symbols present were the untranslated phrases "EN KATU" and "ANKH"—words that had appeared in the seemingly nonsensical speech at the very beginning of Stone's trance episode. Furthermore, among the translated hieroglyphs and utterances were what appeared to be references to the mushroom itself: "The King [or sovereign] speaks of the beautiful sacred plant," "a plant with a red crown."

For all of its consistency, one aspect of Stone's message warrants at least some comment—namely, the discrepancy between the two colors cited in connection with the mushroom: yellow and red. "In Europe and the Far East the *Amanita muscaria* was red with white spots," Puharich observes later in the text, "and in

4   INTRODUCTION TO THE NEW EDITION

the eastern part of North America it was yellow to golden with whitish spots." Notably, the yellow *Amanitas* are specifically cited by the source of the communication in reference to the one that "grows here," that is, in the northeastern United States. Later in Puharich's escapade, a transmission is received that instructs the team not only on *when* the fungus grows in their region ("*July*") but they're actually told *where* it can be found ("*right here in Glen Cove . . . in the woods . . . under oaks . . . on the coast . . . on the Penobscot Bay side [of the peninsula] . . . near the . . . south end of the road*"). And, when they did successfully locate a specimen, that's precisely when and where it was found. I don't want to spoil all the developments in Puharich's plot, however. So, I'll leave the remainder of these terrific twists for readers to discover in their own time.

Puharich's landmark study would go on to influence a number of significant publications concerning the possible role of *A. muscaria* in the various mystery traditions of the ancient world, from ancient Egypt to archaic Greece to Rome in Late Antiquity to Vedic India—all the way up to western Europe in the Middle Ages and Renaissance and even to the pre-Colonial Americas. Indeed, few and far between are the locations where one researcher or another has failed to find at least some evidence for what is interpreted to be *A. muscaria* use within magico-religious contexts. The most notable of these subsequent studies include: *Soma: Divine Mushroom of Immortality* (1968) by R. Gordon Wasson; *The Sacred Mushroom and the Cross* (1970) by John M. Allegro; *Persephone's Quest: Entheogens and the Origins of Religion* (1986) by R. Gordon Wasson, Stella Kramrisch, Jonathan Ott, and Carl A. P. Ruck; *Strange Fruit: Alchemy and Religion: The Hidden Truth* (1995) by Clark Heinrich; *The Apples of Apollo: Pagan and Christian Mysteries of the Eucharist* (2000) by Carl A. P. Ruck, Blaise Daniel Staples, and Clark Heinrich; and *Sacred Mushrooms of the Goddess: Secrets of Eleusis* (2006) by Carl A. P. Ruck. To the present day, similar studies continue to emerge that pick up on the toadstool trail originally blazed by Andrija Puharich. And, if the scope of the available literature on the subject is any indication, we can expect future studies to chronically spring up . . . like mushrooms.

P. D. Newman
Tupelo, Mississippi

P. D. Newman has been immersed in the study and practice of alchemy, hermetism, and theurgy for more than two decades. A member of both the Masonic Fraternity and the Society of Rosicrucians, he lectures internationally and has published articles in many esoteric journals, including *The Scottish Rite Journal, Knights Templar Magazine*, and *Ad Lucem*. He is the author of *Angels in Vermilion: The Philosophers' Stone from Dee to DMT, Alchemically Stoned: The Psychedelic Secret of Freemasonry, Tripping the Trail of Ghosts: Psychedelics and the Afterlife Journey in Native American Mound Cultures*, and *Theurgy: Theory and Practice— The Mysteries of the Ascent to the Divine*.

THE SACRED MUSHROOM

KEY TO THE DOOR OF ETERNITY

ANDRIJA PUHARICH

1959

chapter 1

THERE is an old saying that the best stories are never told. And for several years I kept silent about a story that I participated in to some extent. I felt this way because it was an adventure of the mind that came to me, and, as such, it had a highly private and personal meaning. But gradually I realized that it was not so private and personal as I read books published on different aspects of the same experience.

My experience grew out of observations on a young man who spontaneously went into a deep sleep and then began to speak and write in the ancient Egyptian language. In recording his words and actions there was always the gnawing question as to whether I was observing an elaborate delusion, or witnessing a subtle manifestation of a reality far beyond the boundaries of common-sense experience. While recording this case for three years there came to my attention a number of works, each of which echoed a part of my observations on this young man.

In England there was published the case history of a woman who spoke ancient Egyptian in the trance state over a period of six years.[1] Morey Bernstein published his observations of a hypnotized subject speaking of the memories of a life prior to her date of birth.[2] Each of these published works brought to the attention of the public highly controversial findings, all of which were present in my observations. I admired the courage of these authors and, inspired by their example, finally decided to bring to the public the facts that I had observed.

[1] *Ancient Egypt Speaks*, Hulme & Wood. Rider & Co., London, 1937.
[2] *The Search for Bridey Murphy*, Morey Bernstein. Doubleday, New York, 1956.

10   THE SACRED MUSHROOM

My experience began when I was on duty with the Army of the United States as a medical officer during the years 1953 to 1955 and was stationed at the Army Chemical Center, Edgewood, Maryland. It all began rather abruptly June 17, 1954, when I looked up from my desk to see Sergeant Cairico stop smartly in the doorway.

"Sir, there's a long distance call for you, a Mrs. Bouverie. Will you take it in the other office?"

"Yes, Cairico." I put off my next patient and walked down the corridor. Alice Bouverie was a trustee of a research foundation in Maine where I had been employed before entering the Army. I wondered why she would be calling me now.

"Hello, Captain Puharich speaking."

"Hello, hello, Andrija. I'm so sorry to disturb you during clinic hours, but I simply had to pass this piece of news on to you or burst."

"No disturbance at all, Alice. It is a pleasure to hear your voice after all this clinic clamor."

"Do you remember that Dutch sculptor I introduced you to several months ago in New York?"

"Yes, vaguely. The one who is supposed to be able to describe a hidden picture across a room when he is blindfolded?"

"Yes, that is the man. Well, he and Betty were here last night for dinner. Being a sculptor I thought he'd be interested in some of my pieces. He liked the Nadelman and the Henry Moore, but that's about all. So I thought I'd show him some of the older things and fetched out the Egyptian jewelry. I have a gold pendant[8] that belonged to Queen Tiy—at least Sir Wallis Budge at the British Museum told my mother-in-law that Tiy's name was on it, and that it probably was her personal property. Well, I had no sooner handed it to him than he trembled all over, got a crazy staring look in his eye, staggered around the room a bit, and then fell into a chair. I was petrified and really thought he was having an epileptic fit. Betty said that she had never seen Harry like this before. I rushed to get some water while Betty held him up. When I got back he was sitting rigidly upright in the chair and staring wildly into the distance. He didn't seem to see us at all but was watching something we couldn't see."

"Sounds as though he was in a trance, doesn't it?"

"Yes, that's what it turned out to be, but at that moment I had

[8] In the form of a cartouche.

THE SACRED MUSHROOM 11

no idea what was going on. It probably wouldn't alarm a doctor like you, but I had never seen anyone go into a trance before."

"Well, what happened to make you believe that this was really a trance?"

"He just sat staring into nowhere for about five minutes. Then he jumped up and clutched my hand sort of desperately. I must say, it was awkward and embarrassing, especially because of the way he kept staring into my eyes. I have never seen such fanatically blue eyes in my life. He kept saying, 'Don't you remember me, don't you remember me?' over and over. And I kept saying over and over, 'Of course, Harry, I remember you.' But this made no impression on him. Then he began to speak quite clearly in English about his upbringing. I didn't realize that there was anything extraordinary about what he was saying until he asked for a paper and pencil and began to draw Egyptian hieroglyphs. I'm sure they were hieroglyphs, even though I don't know a word of Egyptian. This finally made me realize that he was in a somnambulistic state. Then he started to tell me about some drug that would stimulate one's psychic faculties. That is why I called you, because you're the only person I know who might be able to make sense out of what Harry said. I'd like to know what this is all about."

"Well, it sounds interesting. Why don't you send me a transcript of what he said, and I'll give you my opinion."

"Fortunately, we took notes. I'll send it special delivery right away. Good-by, and I'll let you get back to your patients."

"Good-by. You'll hear from me as soon as I have something to report."

I walked down the corridor to my office where Colonel Blalock's wife was waiting. My mind was preoccupied with the telephone conversation as I sat down. I scarcely heard Mrs. Blalock as she said:

"Doctor, as I was saying on the telephone this morning, this pain in my back catches me at the most unexpected times and lasts for hours once it starts. And what worries me is that the pain goes around my left side and I feel a pain near my heart also. But Dr. Hopkins took an electrocardiogram a few days ago and says that my heart is O.K. So if it isn't my heart, what is it?"

"Oh yes, Mrs. Blalock. Nurse, will you take Mrs. Blalock into the examining room? I want to examine her chest."

"Yes, Doctor. Come with me, Mrs. Blalock."

While waiting for the preparation to be completed my mind wan-

12   THE SACRED MUSHROOM

dered back to the telephone call. It was odd that I should get such a call in the light of what had happened a few weeks before. Colonel Nolton,[4] Chief of the Army Medical Laboratories of the Chemical Corps had invited me to his office one day. In a most roundabout way he had quizzed me about my experience with mind readers and such people who could get verifiable intelligence in the absence of any known mechanism to account for it.

Now I knew that this was neither a casual nor a fortuitous quizzing. This was only the most recent of a long series of conversations which had started prior to my entry into the Army. The first such conversation had started in August of 1952 at the Round Table Laboratory in Glen Cove, Maine. A friend of mine, an army colonel, who was Chief of the Research Section of the Office of the Chief of Psychological Warfare, had dropped in to say hello. He expressed a rather normal sort of curiosity about my investigations of extrasensory perception and was quite interested in a device which we had been developing in order to increase the power of extrasensory perception. He asked me if it really worked. I told him that I didn't know yet and that I wouldn't for a couple of months, until our statistical analysis of the results of the experiment was completed. The colonel then surprised me by saying that if we found any positive results to be sure to let him know, as the Army was definitely not disinterested in this kind of work.

It was November 1952 before the statistical analysis of the telepathy experiment was completed. The results showed that extrasensory perception was increased in the Faraday Cage[5] device by a healthy margin over those scores obtained under ordinary room conditions. This, to me at least, was an exciting finding, as it represented the culmination of two years of experimentation. My enthusiasm led me to send the results to my colonel friend in the Army. He invited me to give a report on this work at the Pentagon. On November 24, 1952, I made such a report before a meeting of the Research Branch of the Office of the Chief of Psychological Warfare. As far as I could tell at the time my report evoked little interest in this group. (However, I found out much later, after I was in the Army, that I had been requested for active duty in the Army on November 25, 1952.) On December 6, 1952, I received one of those well-known

[4] A pseudonym is here used to protect the privacy of this officer. The privacy of other officers mentioned in this story will be treated in the same manner.
[5] See Appendix 1.

greeting cards from my draft board, took a physical examination, and was inducted into the Army on February 26, 1953.

This induction into the Army had some amusing features. I had had a medical discharge from the Army as a first lieutenant in 1948, and it seemed highly improbable that I would ever see military service again. When I saw that the draft board was serious about getting me into the Army again, I went through the mechanics of applying for a new commission in the Medical Corps about three weeks before my impending induction. The day of induction arrived and I was still without a commission. I called up the head of my draft board and asked him what I was to do. Captain Bard told me to go ahead and report to the induction center and all would be taken care of there.

At the induction center I was whirled through the processing line, and found myself raising my right hand to be sworn in as a private. I objected to this irregularity, but no one would listen to me, even though I was well over the regular draft age. So I bowed to the machinery of the State and found myself a private E-2, of which there is no nobler, nor much lower, at Fort Devens, Massachusetts. I was quite an old man among my fellow inductees, all of whom were in their pimply teens or early twenties. For two days I tried to get the ear of the highest-ranking noncom available to me, one Corporal Blankenberg.

"Corporal," I said, "I'm a doctor and do not belong here. I was once a private in World War II and was an officer as late as 1948."

He looked me over as just one more wise guy and said, "Oh, so yer a docter, hah? I got just the job for you! See dem latrines? Well, we want dem nice and sanitary, and you're in charge!"

My friends on the "outside" finally got the ear of the camp commandant after four days, and a colonel appeared at our barracks to inspect this puzzling medical private. After some apologies, he invited me to ride with him in his car to headquarters. As I got into the colonel's car, my fellow inductees all cheered, while Corporal Blankenberg spluttered in rage. The camp adjutant hastily swore me in as a captain and I was ordered off to the Medical Field Service School at San Antonio, Texas. There was just one catch. My commission had not been activated, and so I was still a private.

On my way to San Antonio I stopped off in Washington to see my Army friends at the Pentagon. They were quite amused at my adventures as a private but told me not to worry, since they wanted to use me in my own "field of research." Thinking that this was some

14 THE SACRED MUSHROOM

sort of Army game that I had best not question, I placidly went off to Texas. However, the game went on, and I remained a private for another thirty days. I was assigned to a class of some two hundred medical officers ranging in rank from first lieutenant to lieutenant-colonel. As a private I aroused great curiosity. My fellow draftees and enlisted men thought I was sent to spy on them because I attended the officers' class. My fellow officers thought I was spying on them because I was a private. Besides, the Korean War was raging full, and everyone was jittery and uncertain as to where he would be sent. So I got no sympathy from anyone, and the Army maintained a stony silence as to why I should remain an enlisted man.

During my first week in San Antonio, I was invited by the commandant of the Air Force Aviation School of Medicine to give a lecture on extrasensory perception. I accepted, but this threw both the local Army and the Air Force commands into quite a tizzy. Finally the two opposing generals agreed to let me, as a private, give this lecture for a group of officers and civilian scientists. Prior to my induction I had given the same lecture in February 1953 at the Pentagon for the Advisory Group on Psychological Warfare and Unconventional Warfare, Department of Defense. I suppose that word had come from Washington to the Air Force Aviation School to request a repeat lecture. Because of the Army and the Air Force interest in this subject, I requested a duty assignment which would allow me to continue my research. I got some whispered assurance to this effect.

My commission was finally activated and I became a gentleman once more. My duty assignment was to the Army Chemical Center in Maryland, which was the hub of quite a bit of psychological, neurophysiological, and chemical research. On the basis of assurances by high-ranking officers I fully expected to be assigned to a research job, since this constituted the bulk of my professional experience. But I found yet another obstacle. I did not have a security clearance. And this would take some time because one senator from Wisconsin had done a good job of paralyzing the judgment of clearance officers, and everyone was playing it safe in those days in the military.

I became the Chief, Outpatient Department, of the base dispensary. All the Army interest in my research work had vanished and I remained just plain "Doc" for the next six months. Because I was inclined to be conscientious I acquired a reputation as the doctor to see when one was ill, and this kept me more than busy, leaving no spare time for my research interests.

THE SACRED MUSHROOM 15

However, in November of 1953 my colonel friend in the Pentagon called me up one day and said that a way had been worked out whereby the Army could sponsor my researches into extrasensory perception. This was to be arranged through a university which would act as a blind for the Army interest in this forbidden subject. After several months of negotiation all this, too, came to naught. Therefore, I was more than surprised to have the subject reopened by a responsible officer, Colonel Nolton, on my own post, however circuitous and indirect the approach may have been. I told him that in my opinion extrasensory perception was a reality, and that it could be proven in people with exceptional talent. I pointed out that there was also evidence to the effect that the talent was widely diffused throughout a normal population, and that it was probable that everyone has some of it sporadically.

"Well, if this is true," he persisted, "isn't it possible to find some drug that will bring out this latent ability so that normal people could turn this thing on and off at will?"

"It would be nice to have such a drug," I replied, "because then the research problems of parapsychology would be half solved. You see, the main problem in extrasensory-perception research is that we never know, even in a person of great talent, when this mysterious faculty will manifest itself. So we just sit around like a fisherman in a boat who puts his hand into the water every once in a while, hoping that a fish will swim into his grasp. There have been some reports of primitive peoples using such drugs extracted from plants, but I have never heard of one that worked when tested in the laboratory."

"Well, if you ever find a drug that works let me know, because this kind of thing would solve a lot of the problems connected with Intelligence." This was the parting word of the colonel as the conversation ended.

"Doctor, the patient is ready."

I examined Mrs. Blalock very carefully and found that she had one of the small muscles in the back near the spinal column in severe spasm. I could find no other cause for her pain. After checking with her previously taken X rays and other laboratory work, I decided that the muscle spasm was probably the cause of the pain. I explained this to her and asked permission to infiltrate the muscle with some novocain. I couldn't promise immediate relief but suggested that if

16   THE SACRED MUSHROOM

this was the cause of the pain she would indeed get relief. She hopefully consented to this trial nerve and muscle block. With the injection of three c.c. of novocain she cried out in exultation.

"Doctor, it's miraculous! The pain is gone, including the pain near my heart. All you did was use that needle on my back, and it didn't even hurt."

I was pleased with this result, because these cases didn't always respond so quickly.

"Please return tomorrow, and I'll check the condition again."

"Thank you again, Dr. Puharich. It sure is a wonderful relief. I pray that it doesn't come on again."

As the day passed and I saw patient after patient, the telephone call of the morning faded from my thoughts. At 5:00 P.M. I checked out of the clinic and picked up my wife and three daughters to go swimming. It was one of those hot and muggy June days in Maryland, and fortunately the post had a refreshing swimming pool. We had several hours of noisy splashing in the pool and then had dinner at a drive-in, which the children considered the greatest of treats. By nine o'clock we were back on the post in our apartment. The children had just quieted down in bed when the doorbell rang. It was a postman with a special-delivery letter.

"Who is the letter from, dear?" asked Jinny.

"Oh, I forgot to tell you. Alice called me up from New York this morning in great excitement, because she thinks she has discovered a new sensitive."[6]

"Who is the sensitive, and what does he or she do?"

"I don't really know what the sensitive does, if sensitivity it be. He is a young sculptor that I met at Alice's several months ago. You know how she collects people, and this was just one of a number of people. She told me that he could tell you what was on a photo if it was hidden across the room and he was blindfolded. He did this for several people that day, and I must say his descriptions were rather vague, although one of them was rather intriguing. I guess he may have some telepathic ability, but I've seen better. Anyway this is a report of what happened last night, and Alice asked me to evaluate it because she feels it may be of interest."

Jinny went back to cleaning up the living room mess that the

[6] A word used to characterize a person who has psychical sensitivity, i.e., shows some form of extrasensory perception.

THE SACRED MUSHROOM   17

children had left, and I sat down in the easy chair to read over the material from Alice. This is what I read:

Dear Andrija,

Here are the notes I took as best I could of what Harry said last night. I hope you can make some sense out of it.

Yours,
ALICE BOUVERIE

June 16, 1954, 219 East 61st Street, New York City

1. "There are also those leaves for treating legs. You[7] wanted to use the mold from wood, but hand far better for softening and shaping. There is a plant from jungle, use the root and leaves—put them in water till soft. The leaves have sharp points and shiny —used for softening bone. KU-FA [seems to be a word for some plant]. Hot damp swampy deep down country from south. They are a black, short, big-skulled people. They brought that man. He was bleeding. We stitched him together with termites, and then thorns outside to mend him. Termites have pincers on heads. I chewed those bitter leaves, taste is ugly, and I put that on wounds after chewing. You could use it now, much more important. Stuff that makes bones soft also—for setting breaks." (Long silence.)

2. "TEHUTI AKH. NESI NESU KHUTA NEFERT KUFA ANKH KHUT. PTAH KHUFU. PTAH KATU. EN KATU."[8] (Long silence.) "I see a hall with a deep opening, and on the left in the opening is a statue.[9] It changes into a person, the person speaks, and then changes back, gets life, moves. I see it many times during the day. I have to have time to meditate long time at night. I know my way so well there—just as well in the dark. It was here you had the drawings, the drawings with the cows and the people with corn in the arms. Why did you take them away? What did you do with the statue? It was there on the right. The base was black and it had a necklace."

3. "My mother and father died when I was little. I was brought up by a man who made buildings—builded it in his mind, and had the others do it as he said. I can see architecture. I was born up there. Your mother writes." (Makes following signature.)

---

[7] This referred to Alice Bouverie.
[8] This is the vocalized ancient Egyptian language. (See pp. 28–30 for translation.)
[9] At this point he sketches a dog-headed statue.

18　THE SACRED MUSHROOM

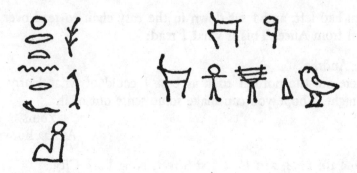

4. "Medicine—light yellow in that jar—of dried wood, it is a powder; very little in my finger; let it fall on that fluid and then it turns red—deep red, that is dry wood. There is a cream to take people out of themselves when they couldn't bear their pain. The same way (be very careful) as a plant I see here. It grows here[10] and works the same way. I don't know how, but I can draw it. There are big ones like this:

The spots are white. The plant is yellow.

"Take the skins off its neck, and the white spots too, they will work the same way. The ointment is to be rubbed in the skull on the joints for trance." (Draws following skull sutures→)

[Indicates a point just above the forehead where this skull suture is located and ointment is to be rubbed in.] "If dose is too much you can get in trance and may never wake up. You are so changed

---
[10] A.B. believes this refers to America.

THE SACRED MUSHROOM   19

in body that you can't get back. Be very careful. Start out with very little and work up. Find what you can stand."

5. "TAHUTI—— I almost remember. I was so close. What are the names? Is that Amenenhotep? I hear connections and names. Bee and honey. Gray pavement. The lake of intrigue. This is getting close. It has something—— Can you remember anything? Do you remember a name? There was a little girl—oh I see a garden coming out on a river, farther back, a building—built of light stone. A little girl—around ten—she hadn't been there long. I'm there too. Remember the name? You were brought there to get the eye and wings. The name you got—when you came there. It is long ago. A boat on the river, its wood and mast is different, it is yellow —an animal painted on the mast:[11] *Antinea*—that name mean anything to you? She has something to do with the temple in the back. There are dark people there too. You learn something like dancing there—to harmonize your body with your soul. An eye to see, and wings to fly beyond. I see a man who brought you. Must be close to you—father—— You ask him to take his beard off, and looked at his face—kisses good-by, and puts beard on. Has something to do with the sun. First thing to learn: people bring all kind of things to temple, God is receiving these things and you have to learn to give it to your people who come on the other side. You have to learn to dance. There came a war later. You learned to write, you draw figures, a bird, flowers. You sit outside by stone, hot from sun. You write in clay, fold and press flat and write again—figures close together."

This was indeed a cryptogram. It was difficult to make a quick judgment as to whether the material represented a fraud of the unconscious, a bit of fancy, or a case of an independent personality communicating through a sensitive until a thorough analysis had been carried out. While in actual fact the analysis of this material went on for several years, it might be helpful to reconstruct the procedure so that the reader, too, may be properly initiated into the fascinating story as it unfolded.

There are several major parts to this communication of June 16, 1954. The third paragraph makes a scanty allusion to ancestry, and

[11] Makes a drawing of an ancient prowed vessel with one large square sail. On the sail he draws a scarab, or beetle.

20   THE SACRED MUSHROOM

because of the hieroglyphs one is led to look to ancient Egyptian sources. But it was the allusion to the plant in the fourth paragraph that first occupied my attention. Here was present a rather clear statement that this drug relieved pain. There was also present possibly the idea that the consciousness could be separated from the body, and that such a separated consciousness could operate independent of the limitations of the body. Now this point formed the beginning of my inquiry. Besides the Army interest, there was another basis for my curiosity, in that two patients had told me separately of having a like experience. Both had been undergoing dental surgery under nitrous oxide anesthesia, one in London and one in New York City, at separate times. Both had had the experience of suddenly awakening and finding themselves, their complete selves, watching the surgeon operate on their physical body. They were completely aware of their personality as they had always known it, and, while the body they watched was their own, it did not seem to belong to them. It was as though they were watching someone else being operated on. When I first heard these tales I was inclined to classify them as rather unusual hallucinations, the kind that a child may have in which he appears to be getting smaller and smaller while the room gets larger and larger. These cases that came to my personal attention had started me some time ago on a search of the literature for other similar cases.

It appears that the ancient Greeks had a tradition, from the earliest recorded times, of men who could detach the psyche, or soul, from the body, travel far, and then return with intelligence from a distance. It is Professor Dodds's[12] opinion that this tradition was not native to the Greeks but that it had come to them from distant Siberia. Here the shamanistic tradition of various tribes, such as the Yakuts, Samoyeds, and Tungus, has carried on the belief that it is possible to dissociate the soul from the body during life, so that each can lead an independent existence for a time.

Professor Dodds says: "A shaman may be described as a psychically unstable person who has received a call to the religious life. As a result of his call he undergoes a period of rigorous training, which commonly involves solitude and fasting and may involve a psychological change of sex. From this religious 'retreat' he emerges with the power, real or assumed, of passing at will into a state of mental dis-

[12] *The Greeks and the Irrational*, Dodds. University of California Press, 1951.

THE SACRED MUSHROOM 21

sociation. In that condition he is not thought, like the Pythia or like a modern medium, to be possessed by an alien spirit; but his own soul is thought to leave its body and travel to distant parts, most often to the spirit world. A shaman may in fact be seen simultaneously in different places; he has the power of bilocation. From these experiences, narrated by him in extempore song, he derives the skill in divination, religious poetry, and magical medicine which makes him socially important. He becomes the repository of a supernormal wisdom."[18]

From this Siberian shamanistic origin the idea of a detachable soul or self spread across the entire Eurasian land mass and became firmly embedded in the stock of Indo-European beliefs. In Greece this idea is associated with the names of Pythagoras, Empedocles, and even Plato. That the idea is still alive is attested to by Professor Hornell Hart's paper in *The Journal of the American Society for Psychical Research*, 1954, in which he describes ninety-nine evidential cases from modern times of this phenomenon which he calls "Extrasensory Perception Projection."

Along with the idea of a detachable soul or self there has persisted the tradition that there were practiced specialized techniques by which this detachment could be achieved in life. I now turned my attention to a search for any description of such techniques in ancient, primitive, or modern cultures. It was important to find out if any such techniques were known, and, if so, was the plant described by Harry known to be used for such purposes. If the plant was known to be used for this purpose one might assume that Harry had fabricated the story in the trance state. If the plant was not known historically to be used for this purpose and if one had some way of testing experimentally the validity of such an effect, then one was confronted with a unique piece of information.

I turned to the material in paragraph four that Alice had sent me. The drawings made by Harry in trance certainly looked like mushrooms to my untrained eye. I must confess that at this time the world of mycology was virtually unknown to me. My only acquaintance with fungi was a half-hour lecture in medical school on the treatment of mushroom poisoning. I knew only that if a patient was brought in and suspected of mushroom poisoning, and that if certain symptoms were present, one used atropine as an antidote. Since

[18] *Ibid.*, p. 140.

22   THE SACRED MUSHROOM

I had never had to treat a case of mushroom poisoning I can hardly say that I was sure of even this bit of information.

The description of the mushroom by drawing was most helpful. I had to look for a yellow mushroom with white spots and one that had a sort of handkerchief around the stem. I was to learn later that this handkerchief was called an annulus by mycologists. The properties of the mushroom, pharmacologically speaking, were to be those of a pain-killer, or analgesic, and probably toxic because of the warning of the danger of death. To me, the alleged psychic effect was to be that of a drug that hastened the detachment of the principle of consciousness from the body during life. With these clues before me I plunged into a world of research rather strange for an Army captain.

I always follow a personal dictum: "When in ignorance, turn to the Encyclopaedia Britannica." Opening Volume 16, I turned to the heading, "Mushroom." Fortunately, there was a beautiful color plate with sixteen mushrooms on it. Of the sixteen mushrooms portrayed two immediately caught my attention. Both were yellowish with white spots on the cap. Both were labeled "deadly poisonous." But only one, the fly agaric (*Amanita muscaria*) had an annulus on the stem that corresponded to the drawing by Harry. I recognized this as the mushroom described in medical school for its dangers to the innocent collector. I decided to pursue this mushroom into the mycologic, pharmacological, and ethnological literature.

It was not long before I found out that the *Amanita muscaria* was the central prop in folklore, fairy tales, and legends. To my surprise there were two distinct colors of the same species. In Europe and the Far East the *Amanita muscaria* was red with white spots, and in the eastern part of North America it was yellow to golden with whitish spots. All the books on mushrooms were in agreement in warning the reader of the deadly poisonous nature of this mushroom if ingested. Yet it seemed that if the mushroom was soaked in vinegar or brine, or that if the skin was peeled off of the cap, it was perfectly safe to eat. In fact, there is some controversy as to whether or not *Amanita muscaria* is indeed poisonous for humans. Traditionally this fungus is noted for its toxic effect on flies, and this is the source of its appellation, *muscaria*, which is derived from the Latin word for a fly. Dr. Smith[14] states that the evidence for the toxicity of the red form is questionable.

[14] *Mushrooms in Their Natural Habitats*, Smith. Sawyer's, Inc., 1949.

THE SACRED MUSHROOM   23

Erich Hesse[15] brought to my attention the fact that the *Amanita muscaria* is used in two widely separated parts of the world as an inebriant and intoxicant. He states that a form of *Amanita muscaria*, called *Nanacatl*, is used in Mexico as an intoxicating beverage by making an extract with milk and mixing it with agave brandy. Reko[16] describes the resulting intoxication: "The effect of the *Amanita* toxin consists of a peculiar hypersensibility. The inebriated person will perceive a mere touch on his skin as highly unpleasant and disturbing. When you blow in their faces they will react with violent gestures of self-defense. The eyes are extremely sensitive to light. Hearing is overstimulated. The sense of smell is changed and they complain of every smell as being unpleasant in an outright pathologic manner. A remarkable symptom is the strong outbreak of perspiration during the first few hours. They also complain of a strong need for urination while they are mostly unable to urinate at all. The psychological behavior is similar to a person heavily drunk on alcohol."

I must state that in the light of later knowledge this description was not exactly accurate in some details, but at the time it was all the information I could find on this subject. The confusion arises from the fact that the observed inebriation is a mixed one due to the combined effects of alcohol and a mushroom. There is some question as to the identification of the mushroom to which Reko ascribes these symptoms. The description of ultrasensitivity of the sensations certainly did not fit the requirements of my cryptogram, which called for a drug that would dull the sensations.

The other area of the world which has had a long tradition of the use of *Amanita muscaria* as an inebriant is eastern Siberia. The association of this geographical area and mushrooms with the previously found shamanistic tradition of separating the soul from the body appeared intriguing. I concentrated a great deal of research on this geographical area, beginning with the earliest known reference to the use of *Amanita muscaria* in Siberia recorded by Philip Johan von Strahlenberg and published in Stockholm in 1730.[17] But nowhere could I find any recorded practice of *Amanita muscaria* being spe-

[15] *Narcotics and Drug Addiction*, Erich Hesse. Philosophical Library, New York, 1946.

[16] *Magische Gifte*, Reko, Stuttgart, 1936, quoted from Hesse.

[17] *An Historical and Geographical Description of the North and Eastern Parts of Europe and Asia, Particularly of Russia, Siberia, and Tartary*, Philip Johan von Strahlenberg. English translation, London, 1736.

24   THE SACRED MUSHROOM

cifically used by shamans for the feat of leaving the body.[18] Western observers had reported only on the use of *Amanita muscaria* as an intoxicant.

My search to date left me more puzzled and intrigued than ever. I had learned a great deal about a mushroom called the fly agaric or *Amanita muscaria*. That it was used as an inebriant in Siberia seemed certain. The evidence to date from the Mexican practice was hazy and doubtful. There was no evidence that the *Amanita muscaria* had any esoteric significance, at least no more than has been attributed to alcohol. The use of the *Amanita muscaria* as a symbol in folk tales, associating it with goblins, fairies, and toads was patently present, but its meaning escaped me. The only clues of any importance I had turned up were the shamanistic tradition of separating the psyche from the body at will, and the Koryak, Samoyed, Yakut, and Tungus practice of *Amanita muscaria* inebriation as having a common geographical origin. If any relation existed between these two practices it certainly was not public knowledge. And in such esoteric matters with their overlay of mystery and secrecy this is just what one might expect. I now turned my attention to what appeared to be Egyptian hieroglyphs in order to further a solution to the puzzle.

[18] *Jesup North Pacific Expedition Series*, Vol. IX, *The Yukaghir and the Yukaghirized Tungus*, Waldemar Jochelson. American Museum of Natural History, New York, 1926. I found out later (1957) that Jochelson had reported such a practice among the Koryaks.

chapter II

Now I had to face a rather unusual problem. Like most people I had never come into any intimate contact with the Egyptian language, especially the ancient Egyptian language. It's true I had been to museums like the Metropolitan, and the Field Museum, and had walked by many mummy cases and sarcophagi looking rather uncomprehendingly at the long rows and columns of pictures that I knew to be Egyptian hieroglyphs. But I can honestly say that I'd never stopped to consider what any particular hieroglyphic picture meant. In other words, Egyptian hieroglyphs were meaningless to me.

There were two questions that I had to resolve in this material that Alice Bouverie had sent. The first was whether I was actually dealing with what could be considered authentic Egyptian hieroglyphs. The second was whether the hieroglyphs, if authentic, were of such a nature that they could easily have been copied and memorized at some time and then regurgitated in a trance condition. I had to settle the entire question of authenticity; in order to make a beginning I, of course, had to go to an expert.

Among my medical friends was a doctor who also had a degree in Egyptology. I turned to him for help. I showed him the material and asked him if he would examine it and tell me what it meant. At this time I did not reveal the source of the hieroglyphs. Being an old friend and rather obliging he went to work on the problem immediately.

He first examined the written material, and as can be seen in the page 18 illustration, on the top there is a circle. Below it there is what appears to be a mouth-shaped line. Under that, a semicircle; under

26   THE SACRED MUSHROOM

that a wavy line; under that another mouth-shaped figure, and under that what appeared to be a serpent or reptile. Alongside this column of figures there was at the top what appeared to be a branch-like representation, and under that a forked stick; below this entire set of figures there was what appeared to be a crude sketch of a sitting figure.

My doctor friend was very patient with me and explained the meaning of each of these figures.

"First," he said, "the circle represents the sound, ḤO. The mouth-shaped figure has the phonetic value, R, and is pronounced RA. The semicircular figure has the phonetic value of either T or TEP and it depends upon the particular context in which it is placed which value it assumes. The wavy line is quite a common and familiar hieroglyphic figure, and it has the value, N. The mouth-shaped figure below that, of course, is R again, and what looks like a reptile figure may represent the letter F. Now the figure that looks like a branch of a tree is the sign of a King and is usually given by the phonetic value of SUT, or NESU. The figure below this is a scepter of a certain nome or principality in Egypt and is known as the Djam scepter. The reclining figure under this entire picture is very difficult to interpret and probably represents somebody who is dead, or the god Ptah, or just any god, although I am not sure about this latter point.

"In reading off the entire set of hieroglyphs," he said, "one does not necessarily read them in the order in which they appear. Especially so, since this may be the name of an individual. If an encircling sign enclosed the entire set of figures it would be called a cartouche. The cartouche was used almost exclusively by kings or nobility. But we do not have such a cartouche encirclement here. However, we do have the branched figure, the SUT or NESU, which means a King. Now, the reading for these hieroglyphs is Ra Ho Tep. The three signs below this could be REN·F which means, this is my name, or it could be pronounced NEFER, or NEFERT. This latter word represents the name of a figure not too well known in the annals of Egyptian history, but is known among scholars because of an unusually beautiful limestone statue that exists of a man called Ra Ho Tep and his wife, Nefert, in the Cairo Museum."

He told me that this man was believed to have lived in the time of the IVth Dynasty of Egypt. The beginning of the IVth Dynasty is dated at approximately 2700 B.C. The date is not certain because of the unresolved Egyptian chronology, but it is an approximate date accepted by most Egyptologists. My friend thought this was a simple and straightforward translating job as far as he was concerned.

He turned to me rather quizzically and said, "Now, why do you bring this to me? If you had looked in a book on Egyptian history you would have run across this particular name."

As he was my friend I thought I had better give him my confidence. I must say I was a little embarrassed in approaching this subject because it put my credulity on trial to tell him that I had an interest in material which was written under a trance state by somebody I didn't really know and had met, only once, and had no means whereby I could vouch for its authenticity. In other words, I might be dealing with a hoax or some sort of hallucination and this would put me in a rather strange position as a physician and, at the moment, as an Army officer.

I took a deep breath and told my friend the story. He was not unfamiliar with this type of phenomena, having written a book along psychiatric lines which covered a similar case, but which was considered more or less as a pathological condition.

He smiled broadly and said, "Now, Andrija, you surely don't believe that this represents the writing of somebody who lived in the IVth Dynasty, do you?"

Somewhat defensively I said, "Well, I have no particular beliefs about this case at the moment; I'm just beginning the investigation. I'm only analyzing this material at the request of a friend."

"Well," he said, "in my opinion I think this material may be an instance of telepathy. Your friend Alice undoubtedly knows some Egyptian, or has seen this particular sign displayed in some museum, and your so-called somnambulistic individual just picked it up by telepathy and wrote it down."

"Well, that's an interesting hypothesis," I replied, "but I think I can say quite firmly that my friend Alice knows little about Egyptian; perhaps a little more than I do, but virtually nothing."

"Well," he said, "in my opinion that's what the case amounts to and nothing more."

There was an embarrassed silence for a moment, as I gathered up my courage and said, "Well, Doctor, won't you please help me to translate one other piece of writing, disregarding the source of the material, and tell me what you think it means?"

"Oh, all right," he said, "that's the least I can do for a friend, but don't take this stuff too seriously." He turned to the hieroglyphs

28 THE SACRED MUSHROOM

which are reproduced on page 16. The top two signs he immediately read as meaning a plant with a red crown. He then read the five signs below.

"Now we see that the first sign looks something like a headpiece, particularly one that might be considered a crown of some sort. The second figure looks roughly like the line sketch of a man. The third figure looks to me like a bowl with two legs on it and some lines crossing the cup part. The fourth figure is a simple triangle, and the fifth figure looks somewhat like a chick."

My friend looked at this writing with somewhat of a gleam of interest in his eye. I could see that now it looked intriguing to him; whereas the previous set of hieroglyphs seemed to be rather common and routine for him.

"Now, there are several rather peculiar characteristics about this writing. Note in the first place this third figure. It looks like a bowl. I want you to know that this is a rather little-used figure in Egyptian hieroglyphics. It seems to have dropped from common use rather early in the history of Egyptian writing. What it actually represents is the fork of a tree with some crossbars placed on it, making a ladder. This sign represents a ladder. Its phonetic value is somewhat debated, but most Egyptologists feel that it represents the value K. Now the figure next to it, the triangle, again looks a little peculiar. Ordinarily, one would say that this has the value of a Q or K sound, but in this particular context again it seems to have a more archaic value; which is not again a certain value, but it may represent the letter T. And, again, this particular value was dropped shortly after the early dynasties. So both of these figures intrigue me in that they are dated for the period of the historical Ra Ho Tep. Now, we turn to the last figure in this row, which is that of the quail chick, and this is a rather common hieroglyph which has the phonetic value, U. Now, these three symbols together—that is, the K, the T, and the U—could represent the word, KATU, and this has the rather common meaning of "the one who is dead"; however, looking at the entire set of figures this appears to me to represent an alternate title. When a prince became the King, that is, ascended the throne—he usually had five names given to him at that time. This might well represent another name of Ra Ho Tep."

He turned to the first two hieroglyphs in this series, and again said that the first one, the one that appeared like a crown, was very in-

THE SACRED MUSHROOM 29

teresting. It was indeed a crown and represented the red crown of lower Egypt, that is, the part toward the Delta where Cairo is now centered.

"However," he said, "this figure is not drawn exactly the way it usually appears in the textbooks; although this is not too important, and it is well known that there were some variations from generation to generation in the representation of certain figures in ancient Egypt. Now, this figure of a crown has the value, N. However, this figure is infrequently used to represent the value, N."

He thumbed through a large Egyptian dictionary and showed me that in the entire fifty or sixty pages covering the letter N the red crown appeared only two or three times as the phonetic value N.

"The next figure," he said, "is a little uncertain. It is hard to tell by the way in which it is drawn whether it is a sketch outline of a man, or whether it is a sketch outline of a common figure which is known as the ANKH. ANKH means, in Egyptian, life, and it is used in many different contexts to indicate the general idea of life. Now, in reading the entire set of hieroglyphs, this possibly could be translated as EN KATU, ANKH, the name being EN KATU, and ANKH being a sort of salutation which means life, or it could be translated as ANKH KU, which means ascension of life."

He looked up at me after this examination and said, "You know, this is rather interesting, and I think I will have to revise my opinion about this being telepathy, because whoever wrote this certainly knew the archaic form of the Egyptian language. Perhaps you had better investigate the background of this sculptor and find out if he, himself, is not somewhat of an expert in this language. As far as I know, there are only a few scholars in the United States who are expert in this particular period of Egyptian history, so-called 'Old Kingdom,' and you might get in touch with one of them. One of them is at Brown University; that is, get in touch with him if this phenomenon continues or if you are convinced that it has some authenticity that merits further investigation."

After this translation in which my friend had come out with the names: Ra Ho Tep, Nefert, a secondary name, En Katu, and a statement about a plant with a red crown as a meaning of the Egyptian hieroglyphs,[1] I took another sheet out of my brief case and said, "Doctor, I'm grateful for your work so far; and, of course, I appre-

[1] See Appendix 2, DRAWING No. 1, for the definitive translation.

30 THE SACRED MUSHROOM

ciate your skepticism because I, too, am skeptical since I don't know anything about this particular case."

I showed him the other piece of paper on which I had copied the second paragraph material from my friend in New York. "The second paragraph contains words which were transcribed phonetically by Alice, and I read them this way: T-E-H-U-T-I A-K-H. The next set of phonetic transcriptions shows that the first word is spelled N-E-S-I; the second one is N-E-S-U; the third one is K-H-U-T-A; the fourth one is N-E-F-E-R-T; the fifth one is K-U-F-A; the sixth one is rather uncertain in that Alice had sent me several spellings. One is A-N-K-H K-H-U-T; another spelling is A-L-G-K-U-T; another spelling is A-R-G-K-U-T. The general phonetic value apparently being A-U-N-G-K-U-T. Now, the next line is P-T-A-H K-H-U-F-U, P-T-A-H K-A-T-U, and the last one is E-N K-A-T-U."

I looked up at my friend. "Look, we have something very interesting here. In the first place, without my giving you these particular phonetic values, you have translated one of these written phrases as 'EN KATU, ANKH,' is that correct?"

"Yes," he replied, "that is correct."

I said, "Now, isn't it rather unusual for a person to write out an Egyptian phrase which you have independently translated as 'EN KATU' and which you told me was written in an unusual form? And isn't it unusual that the same individual should give a phonetic value which closely approximates the written value which you have given me?"

"No," he said, "that is not too unusual. I think you've got a little detective job on your hands, because you either have a case here of somebody who really knows the Egyptian language, consciously or unconsciously, or one who may have been carefully coached in a very unusual form of it. I certainly wouldn't want to have your job," he went on to say, "because trying to get at the truth in a problem like this requires infinite patience and a long, long period of investigation. I wish you luck, my friend, but I must ask you for professional reasons not to bring any more material of this nature to me. I really cannot afford to risk my reputation by dabbling in psychic matters."

"Now, Doctor," I said, "as long as you've gone this far, would you do me one more favor? Please translate the sentence I've just read to you, and I won't bother you any more. I appreciate the delicacy of your position, and I certainly wouldn't want to jeopardize it."

THE SACRED MUSHROOM   31

He sighed heavily and said, "All right, let's take a look at it, but this is the last time. Promise?"

I said, "Yes, I won't bother you any more, Doctor."

He looked at the first two words. " 'TEHUTI AKH,' " he said, "is a rather simple phrase. Tehuti was a great god to the Egyptians. He was known as the god of both writing and mathematics. He is probably among the oldest of Egyptian gods. The Egyptians firmly believed that civilization came to them when a god they call Tehuti appeared and gave them their arts, all their knowledge—their knowledge of building, of music, of writing, of mathematics, of surveying, and so on. The Egyptians have held Tehuti in veneration from the beginning of their history to the very end. In fact, the Greeks were very much impressed by the god Tehuti, whom they called Thoth, and whom they felt was akin to their own god, Hermes, the messenger of the gods.

" 'TEHUTI AKH,' then, as a phrase, would mean 'Tehuti—it is I who speak.' The phrase, 'AKH,' in this instance being found after the name of a god or King who is speaking usually means, It is I who speak, the AKH here being different from the ANKH which I have previously discussed. The phrase 'Nesi nesu' is well known to Egyptologists.

"Now, the phrase, 'NESI NESU KHUTA NEFERT KUFA,' I would translate roughly in this manner: 'The King [or sovereign] speaks of the beautiful sacred plant.' This is what I would get offhand from this phrase."

He went on to say, "The last word,[2] whose phonetic value is not quite clear, that is, whether it's ANKH KHUT, ALGKUT, or ARGKUT, would offer some difficulty, but in view of the context in which it is placed I would choose the phrase 'ANKH KHUT' which could be roughly translated as 'plant of life' or 'tree of life.' So the entire phrase would read, 'Tehuti, it is I who speak, the King, of the beautiful sacred plant of life.' Now, does this mean anything to you?"

I again brought out one of my sheets from the brief case, which I have labeled Paragraph 4 previously, and showed him the drawing of a mushroom which Harry had drawn, and which he said had very unusual properties: one, that of a pain-relieving drug; and, two, that of a drug which made it possible to get the principle of consciousness out of the body during life. His speech also indicated that

2 Later evidence revealed the correct spelling to be AAKHUT.

32   THE SACRED MUSHROOM

there was some great danger attached to this operation, and that it should not be gone into lightly.

"Well," he said, "this case gets more and more puzzling and intriguing, but remember that when you walk out of here I won't be connected with this stuff."

I could understand why he felt this way, but I persisted. "Would you please go ahead and translate the last few phrases that are present on this paper?"

"Yes," he said, "I'll do that. Now, you have the phrase, 'PTAH KHUFU,' which again appears to me to be very interesting. You see, it is believed that Ra Ho Tep, as a historical personality, lived at about the time of a Pharaoh who was called Cheops by the Greeks, and was known as Khufu in his own time. Now, it may be that this phrase 'PTAH KHUFU' sort of historically places Ra Ho Tep. Now, this would be very interesting because it would confirm everything that I have said so far regarding the period in which this writing was used and the time in which Ra Ho Tep lived. Historically, it is not exactly certain whether Snefru was the first King of the IVth Dynasty or the last King of the IIIrd Dynasty. Khufu is believed to be either the first or the second King of the IVth Dynasty; that is, Snefru being first and Khufu second. Now, it is believed that there is some relationship between Snefru and Ra Ho Tep because Ra Ho Tep's tomb is found adjacent to that of the pyramid tomb believed to be that of King Snefru, or an earlier King, Huni. There is a further question as to Ra Ho Tep's relation, in that in Snefru's list of titles Ra Ho Tep is listed as being a son of the King. This may mean that he was the legal son of Snefru, or it may mean that he was a son of one of Snefru's other wives. In any case, the historical question is unsettled. With the phrase 'PTAH KHUFU' we have a signature which is more precise than any given so far as to the period of your sculptor's Ra Ho Tep. If you will pardon my laxity in saying this, in the light of what I have said previously, it may be that Ra Ho Tep did live in the time of the Pharaoh, Khufu.

"Now, the phrases 'PTAH KATU' and 'EN KATU,' I believe, refer to other titles of Ra Ho Tep. It was indicated in the previous hieroglyphs as another name; that is, En Katu. If this be true, then Ptah Katu would be another variation of the same name, but indicating a more exalted station by virtue of the use of the title, Ptah."

"Doctor," I said, "I'm very grateful for all this information, but it

THE SACRED MUSHROOM   33

raises more questions than it answers. You see, I know nothing about Egyptian history. If you don't mind, I'd like to ask you a few questions. In the first place, what is a dynasty in Egyptian history?"

The doctor explained. "Well, a dynasty is a series of Kings that were more or less connected by a direct line of succession, and when that succession was broken, a new dynasty began. It does not necessarily mean that a dynasty contains one continuous bloodline, but it more or less means that a continuous succession was in order. There does seem to be some break when Khufu or Snefru reigned. It is suspected that Cheops or Khufu was more or less a usurper in that he did not fit into the direct bloodline. It is known historically that he did take under his protection the wife of King Snefru.[8] By such an arrangement perhaps he was able to confirm his right to the throne."

"Well, that clears up one difficulty, Doctor," I said. "I have another question. It seems to me that I remember vaguely somewhere that Cheops was the man who built the Great Pyramid. Is this true?"

"Yes," the doctor replied, "Cheops or Khufu is the man who built the Great Pyramid at Giza which is the greatest of all the pyramids. Now it is known historically that it took about twenty years to build this pyramid. Cheops or Khufu—I prefer to use the name Khufu because it is more accurate—got every able-bodied man in Egypt to contribute three months of his time a year to this gigantic building project until it was finished. The Great Pyramid at Giza is still one of the greatest marvels of engineering construction ever attempted. It has never been duplicated for size and amount of material moved in one structure. As you know, there is a great deal of speculation about the meaning and the mystery of the Great Pyramid. There are all sorts of ideas about the meaning of the angles of the passageways and the various features of its construction. One of its unique features is that there is a shaft on the south face of the Great Pyramid which is so arranged that on a certain day of each year, which is the beginning of the Egyptian year, the star Sirius—deified as the god Sept—on its rising would shine into the eye of the dead Pharaoh

[8] One authority states that the wife of King Snefru, Queen Hetepheres, was the mother of Khufu. *The Pyramids of Egypt*, I. E. S. Edwards, Penguin Books, Baltimore, Md., 1947, p. 102.

Petrie states that "The intermarriage with the high priests of Heliopolis appears to have begun under Seneferu, the first king with a cartouche. His son, Ra Ho Tep, was high priest (*Medum*, XIII), while his daughter conveyed the kingdom to Khufu." *The Royal Tombs of the First Dynasty*, Sir W. M. Flinders Petrie, 1900, Vol. I, p. 37.

34 THE SACRED MUSHROOM

down this long passageway which ended in the interior of the King's chamber."

"Well, Doctor, all this is very intriguing; and I'd like, for my own benefit, to recapitulate what you said. Now, if I understand you, these hieroglyphic inscriptions refer to a man called Ra Ho Tep, and it is indicated that the wife's name is Nefert. Is that correct?"

"Yes," the doctor said.

"I also gather that another name of Ra Ho Tep was En Katu from the hieroglyphs and from the verbal statement, Ptah Katu."

The doctor said, "Yes, as near as I can ascertain, this is correct."

"Now, if I understand you, these inscriptions show certain archaic features which place them in the oldest period of Egyptian history, that is, the Old Kingdom, and, more specifically, a historical Ra Ho Tep seems to have lived about 2700 B.C., making his lifetime about forty-six hundred years ago?"

"Yes," the doctor said, "as closely as we can ascertain, this is correct."

"Furthermore, it seems that in the verbal statement, it is Tehuti who is speaking, because he says, 'Tehuti AKH,' and Tehuti AKH seems to mean, 'It is I, Tehuti, who speak.' Is this your impression?"

"Yes," the doctor replied, "it would seem that in the instance of the hieroglyphic writing we have Ra Ho Tep represented. In the case of the direct verbal statement it is not Ra Ho Tep who is responsible for the Egyptian words, but it is Tehuti—Tehuti, the god of writing and mathematics of the Egyptians, who is speaking. And if I am correct in my interpretation," the doctor continued, "Tehuti wants to make some sort of announcement of what he calls the plant of life, and which may be further described by the phrase, plant with a red crown. I might say that in almost all the ancient cultures—the Chinese, the Indian, the Sumerian, the Babylonian, the Syrian, and the Egyptian—there is reference to the so-called tree of life, or as the Chinese call it, the plant of immortality. It seems there was a belief in ancient times that there was some sort of plant which bestowed the gift of immortality on him who should partake of its fruits. It would be very interesting, Andrija, for you to follow up this particular line in the different countries I have cited and see if you cannot find some common basis for this belief in the plant of immortality. You seem to have a sort of a clue in this mushroom phenomenon. I, myself, have never heard of a mushroom being given the attribute of a plant of immortality, and it is a completely new idea to me. But since it is a new idea it might be well to explore it to the very end and see where it leads."

I was somewhat bewildered by all the information that the doctor had given me. I thanked him for his valuable translation and for his fatherly advice. As I walked out of his office and got into my car to drive back to the post at Edgewood, my mind was racing with thoughts and questions about what had just happened. My previous studies had indicated to me that the written material Alice had sent had reference to two main topics; the first being a plant that had anesthetic properties and which seemed to have the property of dissociating the principle of consciousness from the body so that the principle of consciousness, or soul, if one wanted to call it such, could operate independently of the body. Furthermore, in my inquiries into the mushroom phenomenon I had found that there was in existence historically some sort of a mushroom cult, but one whose prime purpose was to give inebriation or intoxication to the individual who used it. It seemed that both the phenomena of dissociation of consciousness from the body and the mushroom intoxication had a common geographical origin in Siberia. Now, with the information that my friend had supplied about the rather veridical nature of the Egyptian hieroglyphic writing, and what appeared to be vocalized Egyptian, there was a reference to a plant, or tree of life, with its hint of a mushroom phenomenon existing in the Egypt of about forty-six hundred years ago.

What was the connection between the Siberian shamanistic phenomenon of leaving the body and mushroom intoxication, and the newly suggestive idea that such a phenomenon was practiced in Egypt by means of a mushroom? Could the red crown plant be a red mushroom? The earliest historical references to the former phenomenon in the Western world comes from ancient Greece, and this goes back only to about the year 500 B.C. Here, however, there seemed to be indications that somebody in Egypt, if all this was entirely bona fide, had existed and practiced it almost two thousand years before its historical appearance in the Western world of Greece. I now knew that I had a great deal of historical research before me, and also had to be most careful in assessing the motives of Harry Stone, who had imparted the present information to Alice Bouverie.

chapter III

I PHONED Alice in New York, and reported my findings on the cryptogram she had sent to me. I pointed out that the hieroglyphs were indeed Egyptian but that they raised the question as to how Harry had come to write them. Did they come out of his own personal experience? Was this some sort of a fraud? Or was he showing an unusual form of sensitivity? These questions called for a further investigation. I emphasized my interest in the description of the mushroom usage, and hoped that she could get further information for me. This meant that she would have to see more of him, and I cautioned her about future dealings in these words:

"I would suggest, Alice, that you go ahead and see this sculptor when it is convenient and rather easy to do so. In other words, I don't want you to go out of your way to encourage him.

"Since you are a person of some standing, it may well be that he has concocted some sort of a scheme to get you involved in these ideas and perhaps work some sort of a swindle. Now I hesitate to use the word 'swindle' but as you know in these psychic matters one always has to be careful, because the obvious thing that one runs into is a confidence game."

"Oh yes, I understand what you mean," replied Alice. "I will be very, very careful in dealing with him. It is rather easy for me to invite him out to dinner or suggest or arrange some sort of an art showing, and this will give us some common ground on which we can get together."

"That's quite the thing to do," I said, "and it is my feeling that if there is any genuine basis to this phenomenon it will happen again.

37

38 THE SACRED MUSHROOM

By this I mean that if you see him, please don't ask him to do a show for you. This would be the wrong approach. The best thing to do is to be with him quietly, and talk about things in general, and if this thing is real it will occur of its own accord. You won't have to encourage it. Do you understand what I mean?"

"Yes, I understand you. I will try to see him casually and see what happens, and I'll let you know what occurs."

Any further ideas I may have had about studying the material Alice had sent me were completely dissipated by the heavy demands of my medical duties. The post was in the midst of what appeared to be the threat of a polio epidemic, and I had my hands full day and night, mostly in screening out potential polio victims.

But this was not my heaviest responsibility, strange as it may seem. About this time there was a resurgence of interest on the part of the research personnel on the post concerning the use of psychochemicals which had been brought up earlier by Colonel Nolton.

While this is not properly a part of this story, I include it here to show why my interest in the neurochemical and psychochemical aspects of the human nervous system was stimulated. I had talks with the Chief of the Neuro-Physiology Laboratory, the Chief of the Radio-Biological Laboratory, and the Chief of the Army Medical Laboratory. Long discussions were held concerning the possibility of finding some drug that would stimulate latent extrasensory perception in human beings. During the time I was in medical school I had spent three years of my spare time in doing research work on neurophysiological problems. In order to keep up with these many specialists I had to do a great deal of night work in reviewing my experience in this field.

It was quite clear that nobody on the post had any idea as to the existence of any drug that could guarantee the appearance of extrasensory perception in normal individuals. Now, in the material that Alice had sent me, there was a clue that a certain mushroom might perform this function. However, it would have been sheer folly for me to pass such unverifiable material on to the military authorities. Therefore I devoted quite a bit of time investigating this problem on my own.

I wrote to the Boston Mycological Society to find out if there were any places in New England, which was my residence, where mushrooms fitting the *Amanita muscaria* description could be found. They wrote back and said that some of their members in the past had infrequently reported seeing such a mushroom, but that there was no way of predicting where the mushroom could be found since it never seemed to appear twice in the same place in succeeding years.

During weekends I took my wife and children on long walks through the woods in the vicinity of the post on which I lived, and we looked for mushrooms. However, I must say that during the period of July, August, and September of 1954 we never found a mushroom that fitted the description of the *Amanita muscaria*.

The talks with the military can be briefly summarized: I had not as yet obtained a security clearance even though I had spent seventeen months in the Army. The military authorities, while keenly interested in my past experience with the problems of extrasensory perception, were reluctant to discuss matters completely and fully for security reasons. Therefore, I had to infer a great deal from what they said as to their real interest. It was my conclusion that there was no working program in operation as yet in the field of extrasensory perception, but that I was involved in what might be called an exploratory situation at the moment.

The commanding officer of the post seemed to take a great personal interest in this particular development. He and his wife both spent some time with me, informally discussing this problem. I must say that their interest was a skeptical one, but nevertheless a real one. I kept up as best as I could with my medical work, these discussions with the military authorities, and with my family life.

In the middle of August I received word from the post security officer that my security clearance had been granted. This was most welcome news to me and ended a period of long anxiety and uncertainty. Now I was, in the eyes of the Army, ready to receive more detailed information about the psychochemical research program which had been under discussion for many months.

The upshot of many and long conferences about this subject was a proposal from the commandant of the post. He explained to me that it was very difficult for the military to engage in the kind of research we had been discussing. The main reason for this was that anybody inside the military who was interested in this subject was automatically branded as a crackpot. Furthermore, the subject had certain unfortunate political implications which could react unfavorably on any individual who put them forth. The general proposed that it was possible eventually, if I was patient, to continue my two

40   THE SACRED MUSHROOM

prime interests within the military—that is, my medical work and researches in extrasensory perception. However, in order to do this it was his suggestion that I apply for a commission in the regular Army rather than the reserve commission which I now held. This would insure the Army a firmer grip on my future, and if I proved myself worthy he felt that there would be no obstacle to continuing my two main interests.

I was quite interested in this proposal at the time but certainly could not feel free or ready to accept it. I explained to the general that I had spent many years in building up a laboratory in Maine, and that there were many people and my staff who were counting on my return. I pointed out to him that my loyalty to them came first, but that loyalty to the best interests of my country was also in my heart. I asked him for time to consider the proposal.

While these conferences were going on with the research and military authorities I had been receiving more material from Alice Bouverie in New York where she had apparently had several meetings with Harry Stone. During each meeting he had unexpectedly gone into a trance. I was much too busy to study this material in any great detail, but I could see that there was more Egyptian writing, more vocalized Egyptian, and a great deal of information about Ra Ho Tep in the English language.

Although I looked at the material very casually, my interest in this case heightened. I realized that in order to understand it I would have to go to New York and interview the Dutch sculptor, and, if possible, see him at work in order to make a personal appraisal of the phenomenon. Therefore I arranged with Alice to visit her in New York on the weekend of September 4.

The material which she had already sent me I must reserve for presentation to the reader in a later chapter, since at this time it was meaningless to me. I might say that my experience had been very disheartening in trying to interest Egyptologists in the material already at hand. As soon as I mentioned the word "medium" or "psychic" they immediately lost interest.

I arrived in New York early in the afternoon and went directly to Alice's house on East Sixty-first Street. I brought with me a tape recorder and some other test materials for any eventuality that might occur. I placed the tape recorder in the study in an inconspicuous position so that if Harry did go into a trance or begin to speak Egyptian I could record anything he said.

THE SACRED MUSHROOM   41

In the meantime Alice had arranged for Harry and Betty to come over for tea. They arrived promptly at five o'clock. Harry Stone was a small slender young man with bright blue eyes, dark blond hair, and a full beard. I must say that I was rather surprised at the full beard, because he didn't have one at our first meeting many months ago. I understood that Harry was living a sort of garret life as a sculptor in New York and working part of the time in a mannequin factory making plaster models in order to support himself.

When Harry greeted me, he shook my hand with one nervous pump and then shyly retreated into himself. I found him to be extremely sensitive to people's feelings but highly introverted at times. He seemed almost abnormally shy; however, when confronted with a good flow of conversation in a sympathetic vein he loosened and began to talk about himself. He told me he had come to the United States from Holland about three years ago. He had had a high-school education in Holland and had gone to an art academy for several years. He had been trying to establish himself for the last six or seven years as a sculptor with some modest success. I inquired more closely as to his experience in matters psychic. He was quite forthright in stating that he had from his earliest childhood memories always lived in some sort of a daydream world of his own. For example, as a small boy he had invisible companions with whom he played. He also felt sure that an old kindly man had appeared to him occasionally who would speak to him, and he would speak back. It must be emphasized that all these creatures were imaginary in the sense that nobody else saw them, but to Harry they were quite real.

One day when Harry was about six years old he suddenly announced to his mother that a favorite aunt was dying at the moment in the hospital and would be dead shortly. The mother sternly upbraided him for such a wild suggestion and put him to bed. That evening, about six hours later, the family received word that the aunt had been struck by a car and had died in a hospital. After this the parents became quite concerned about their son, Harry. Being devout Catholics, they, of course, went to the parish priest for advice.

Because of this experience and other like experiences, Harry says he was lectured to many times by his parish priest about the desirability of avoiding them since this clearly was the temptation of the devil. Over the years Harry had had instilled into him the idea that any such experience was evil and bad for one's soul, and was to be

42   THE SACRED MUSHROOM

avoided at all costs. Occasionally he would show flashes of keen intuition which could only be explained on the grounds of telepathy or some other form of extrasensory perception. And this would always occur in embarrassing situations where he would show foreknowledge, and of which he should have had no knowledge. In general, his life was traumatic for this reason. There was little question about it. He was very embarrassed to speak about it and wished, as he said over and over again, that *it* would leave him alone. In fact, he said, when he emigrated to the United States, he had sworn to himself that he would never speak about these things or reveal them to anyone, because this was definitely not an asset in his life.

But one day he was attending a party of some painters in Greenwich Village where one of the artists expounded at great length as to all this "psychic bunk" and all this "ESP[1] hooey." This so angered Harry that he challenged the man to test him, and offered to show him that ESP indeed was true.

The man asked, "What kind of a test can you do?"

Harry said, "Well, if you will just take a photograph out of your pocket, and stand across the room with your back to me, I will tell you what is on that photograph." According to Harry, he described that picture completely. The man was horrified that such a thing could happen to him. This was the first time that Harry had revealed his abilities in this country, and after that there was only one other demonstration of his abilities until the time he had come to visit Alice.

About two weeks before this "thing" had occurred in Alice's presence, Harry said that he had been sleeping very badly and had had terrible nightmares and a severe pain in his back. After about a week of this he decided to go to a doctor. The doctor kept him under observation for a week, cleared up his back pain, and told him that there was nothing the matter with him and to forget about it. He was visiting Alice, whom he had recently met at a party, and was just enjoying a conversation about sculpture when she had handed him the gold cartouche. He said that the moment the cartouche touched his hand he felt as though an electric shock had gone through his whole body and he remembered nothing after that until he woke up and Alice and Betty had described to him what had happened.

Since this initial experience he had been free of his nightmares and insomnia and had slept quite well, but it seems that every time he

[1] Extrasensory Perception.

THE SACRED MUSHROOM 43

came to see Alice he got this incontrollable desire to go to sleep; and to sleep he would go. When he woke up, there would be his friends sitting around telling him that he had said certain things when he was asleep. It never happened anywhere else. This was so embarrassing that he just didn't know what to do about it, and he apologized profusely to Alice for having done it and hoped it would never happen again. I told him not to be concerned about it since these things were known, and that there was nothing dangerous about them, if properly conducted. This seemed to be reassuring to him, but I could see that he was still very uneasy about the whole situation and, in fact, still embarrassed that it should exist at all.

After tea Alice took us out to dinner, and we had a jolly time talking about things of mutual interest. After dinner we returned to Alice's house and sat down in the study to have some coffee. While the maid was serving I took Alice aside and asked her if she would please lend me the gold cartouche so that I could carry it in my pocket. I told her that sometimes an object like that acted as an inductor and set people off in these states if it was a genuine phenomenon; and I said if the opportunity presented itself I would do that tonight. She assented and gave me the cartouche.

I returned to the study and sat down next to Harry and again began asking him about his psychic experiences. While he was telling me about the various spontaneous things that had occurred in his life, I asked him if he would be so kind as to do a little demonstration for me. He said he wouldn't mind although, if possible, he would rather not do it. I told him that I was in town only for the day and I would be most interested to see what he could do.

He finally agreed and allowed me to place a secure blindfold over his eyes. Now Harry was under the impression that I was going to take out a picture, and have him read it as he had done for other people. But this was not my intention at all. I wanted to see if this gold cartouche had any special power which set Harry off into his alleged trance condition. I had seen many sensitives and mediums go into the trance state and felt competent to judge whether I was witnessing a genuine or an assumed trance. Harry sat down on the couch opposite a coffee table after I blindfolded him, and I sat next to him.

I did indeed give him a photograph, which I allowed him to hold in his hand because I had already placed it in a sealed envelope. He held this for about five minutes and concentrated and then said that it meant nothing to him. He could get no information that would be meaningful. He again was quite embarrassed about it.

"Now," I said, "Harry, don't be upset about this, please; because these things don't always work on demand, and I have made a sort of demand on you, and I am not at all disappointed. Would you mind trying one more thing which I have with me?"

He said no. He would be willing to try, but he had no confidence in himself. While he was blindfolded, I placed in his hand the small gold cartouche wrapped in cotton wadding which Alice had used on the evening of June 16 and which apparently was associated with his going into a deep trance. I was quite unprepared for his sudden reaction. His whole body stiffened; his head became erect; he breathed irregularly in long gasps; his limbs trembled. Then his tension slowly relaxed and he slumped back in the chair. He began to breathe very heavily, sighing as he breathed. His body began to sway to and fro. He raised his arms and swept them back and forth in front of him. He began to try to articulate, but all that came out of his throat were gurgling noises. This went on for about five minutes. It certainly was not the polished performance of a practiced medium, genuine or fraudulent. He seemed to be struggling with himself or with something. It was hard to tell what it was that he was trying to overcome.

But he finally waved his hands around with a motion indicating that he wanted a pencil. I placed a pencil in his right hand and a sheet of paper under it. I watched him closely. I saw that his blindfold was still in good condition and that he could not easily peek under it. He then reached over to the paper with his left hand and held it firm. With his right hand he began to draw. It interested me that, as he began to draw, his head was not inclined at all in the direction of the paper, but was slumped down on his chest. There seemed to be no relationship between his line of vision as indicated by his slumped head position and the part of the paper on which he was writing. To my surprise he smoothly and easily wrote down a series of hieroglyphic characters even though he was blindfolded. I have never seen anyone do anything like this before. The line that he wrote is reproduced here.

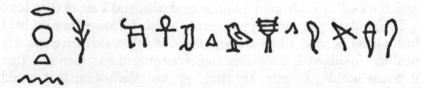

THE SACRED MUSHROOM   45

It was several years before this material was translated. At the time it meant nothing to me. Because I could not get an Egyptologist to translate it, it remained a mystery to me. But as the years went by and I gained in proficiency with this language, I was eventually able to translate it.[2]

Then Harry leaned back in his chair still in trance and began to talk in English. His voice sounded pretty much as it did when he was awake and talking. There was no attempt at dramatizing what he said or placing any foreign inflection on the words. While his own English was tainted with a Dutch accent, it nevertheless was good English. The following is a transcript of what Harry said at this time:

September 4, 1954.

PERSONNAE: A. P. *Andrija Puharich*
          A. B. *Alice Bouverie*
          H. S. *Harry Stone*

A. P. *"Do you know what that says, Harry?" (In reference to hiero-glyphics.) (H. S. does not answer this question.)*

A. B. *Will you talk with Andrija? He would like to talk to you.*

H. S. *She helped me!*

A. P. *Who helped you?*

H. S. *Antinea.*

A. P. *Antinea—she helped you?*

H. S. *She was very handy when we made this. She always wanted to be too fast. And when we made, you know, that bag—when we put it on that leg to make the bone soft. You know, she wanted to hold the bone in wood so that it wouldn't bend any more. I told her we had to do it with the hands, and many times. She thought we could do it at once. But that would flatten the joints when there is too much pressure; you see? And the skin gets very sensitive, it looks like the skin of an elephant, but that goes away after.*

A. B. *Do you think that would be useful now?*

H. S. *Yes, to make the bones soft. It is still there.*

---

[2] See Appendix 2, DRAWING No. 2, for the detailed translation. This phrase means:

A friend of the King. A friend returns [with] the red crown of ascension [over] life.

46 THE SACRED MUSHROOM

A. B. *You once talked about the mushrooms, the yellow ones with the white spots.*

H. S. (*Interrupts* A. B. *here.* H. S. *is very agitated.*) *Why did they take it away? Why did they take it away?* (*Long pause.* H. S. *is breathing very hard and shows a gasping, weeping kind of agitation and struggles with himself.*)
(*The bandage is taken off his head.*)
(*Then* H. S. *places his right fist under his chin—the beard sign of promising something.*)

H. S. *Does that mean anything to you when you promise something?*
(H. S. *looks pleadingly at* A. B. *It is not clear whether he is reminding her of a promise she made to him about keeping the knowledge of the mushroom secret, or whether he is under promise not to say anything more about the mushroom.*)

A. B. *Yes, I know it is a way of promising.*

H. S. *You must remember, remember——* (*Long pause.*)
S ANKH NB HE [R?] FA N PTAH KHUFU ANTINEA.
(*The translation of this vocal Egyptian phrase is given, although at the time of the recording its meaning was unknown.*)
TRANSLATION: "The Lord of Life enfolds [in protection] Ptah Khufu and Antinea."
(*It was clear that* H. S. *referred to* A. B. *as Antinea. No reference was given as to the identity of Ptah Khufu.*)

H. S. AAKHUT AIY RA. KATUHOTEP. AMENHOTEP.
TRANSLATION: Hail, to Ra of the Horizon!
Katuhotep. Amenhotep.
(*The latter two names seem to refer to the speaker himself, and may be interpreted as titles whose meaning is obscure.*)

H. S. *People were wounded in war. Just like the old man, and he had much pain. I brought him—to take him out of his body so that he could rest, and then we could put it together inside. Then we poured that. I see myself, having in a glass something like water, but there is a little bit of powder from wood, and that falls in there, and then it turns red. And that's what I pour inside over the wound.*

A. P. *And that takes the pain away?*

H. S. *He has no pain, because I took him out of his body for a time.*

A. P. *How long was he out of his body?*

H. S. *Until he could stand it again.*

THE SACRED MUSHROOM 47

A. P. *That was a great thing to do for someone in pain from a wound.*

H. S. *Is that normal? Don't they do that now?*

A. P. *Yes, we do that now, except we don't take them out of their body. We don't know how to do that, Harry, but we give them medicines that get rid of pain.*
Harry, who was the King of your land?

H. S. *Ra Ho Tep. Me. (Long silence. H. S. blindfolded again.)*

A. B. *You reminded me once of being able to travel in thought, rather than physically being able to go to a country. You said you took a plant, and then left the country. You were speaking the other day about various plants, and one day you talked about some mushroom that you had, to make some medicine to help people when they were in pain and that would put them out of their bodies. You drew some little plants, some little mushrooms, plants with little white spots, and I remember you describing that, and I would like to know more about it.*

H. S. *Pulled the skin. High trees. (H. S. starts to rub vigorously on the wood of the table in front of him. He no longer speaks, but tries to convey his ideas by gestures.) We rubbed it.*

A. P. *You rubbed the wood and the oil came out? (H. S. shakes his head no.)*

H. S. *Rubbed the plants.*

A. B. *Ground them up?*

H. S. *(Grunts with assent.)*

A. P. *Is your name Ptah Katu?*

H. S. *Me.*

A. P. *Ptah Katu, when you rubbed the plants, the white spots, did you mix them with something? (H. S. again indicated by a gesture something from a high tree.)*

A. P. *Tall trees? There was something that came from the tall trees? Palm trees? Coconut? (H. S. nods in assent and now begins to rub the top of his head.)*

A. P. *Did you use it like that on the head? Did you use it only on your sick people? (H. S. shakes his head no.)*

A. P. *On the priests? (H. S. nods yes.)*

A. P. *Is that how Antinea was initiated? (H. S. nods yes.)*

A. P. *How was this done?*

H. S. *By opening the door.*
By stepping in.

48 THE SACRED MUSHROOM

And by leaving.

But it was only for them who know.

It would be dangerous to say everything one knows. Isn't it?

A. P. *If the mind left the body, Ptah Katu, did it travel?*

H. S. *(There is a long pause.) You speak so strange.*

A. P. *Did you go beyond the land of Egypt to the dark people?*

H. S. *I went. We—— (Silence. Indicates by gesture some far-off place.) But I—— Once I took you by the hand, and I—we went to the sun.*

A. P. *Oh, that's what he meant—where the sun goes down, the horizon.*

H. S. *To the West. You know that terrible noise? We couldn't stand it any more. Terrible light. You said—there was nothing that could frighten you. So we went. And then we came back. We heard a tremendous noise what the sun makes.*

A. P. *What made the noise?*

H. S. *The sun.*

A. P. *The sun?*

H. S. *I could feel close, but we didn't go closer than we could. It is such a terrible sight! Such a terrible noise! Such a terrible cloud! Such a speed! Things happened. They could look so far. We saw many things. And, that is, not ourselves.*

A. B. *Things that were not like yourselves? Or people that were not like yourselves? Or things that happened many, many years later?*

H. S. *But they are secrets. If I would tell the people that our form of religion once couldn't be there any more, that would confuse them too much. (H. S. refuses to say any more on this subject.)*

(H. S. lapses into a long silence. Before this session A. P. had prepared a tray of clay one inch thick and ten by fourteen inches in size. A. P. placed this tray on H. S.'s lap, and handed him a hardwood modeling tool. H. S. picked up the modeling tool and tried to place it in his right hand. It didn't suit his grasp, and so he gestured in a request for a knife. A. P. handed him a sharp pocket knife. It must be remembered that H. S. was blindfolded. H. S. swiftly and surely carved the modeling tool in certain curves that would fit the index and the third finger when it was held as a stylus. It was terrifying to see him handle the sharp knife blindfolded while carving the very hard wood. Some time later this newly carved stylus was taken to the Metropolitan Museum and found to be an excel-

lent copy of an ancient Egyptian clay-writing stylus. H. S. then inscribed with the stylus the following hieroglyphs on the clay tablet:

TRANSLATION: The supernatural power of [a pair] of
AAKHUT. Ra Ho Tep.

(See Appendix 2, DRAWING No. 8, for translation details.)

* * * *

The trance was ended. Gone was the agitated tension and the struggle. Harry quietly passed into a deep relaxed sleep before our eyes. Alice looked at me and said:
"What do you think of it now that you've seen it?"
"I don't think he's faking the trance; it seems to be the real thing. But what does it all mean—this Egyptian language, and the mystery about the mushroom? If we can get this language decoded we may get an answer. He doesn't say much of importance in English, and retreats from direct questions about the mushroom. I feel that Harry is under a powerful influence when he writes in hieroglyphic. He appears to be a machine responding to a master control. I wonder if he could be under intelligent control?"
Mediumship is still one of the little understood aspects of human behavior. There is a compulsion on the part of some human beings to enter into this experience, and to enjoy the drama of slipping into a childlike dreamy state and speaking uninhibitedly of the strays and wisps of thought that drift up to consciousness. Others as spectators enjoy this cloistered drama in a shaded room in the belief that they are listening to the thoughts, voice, and feelings of the departed dead. Like all drama there must be the actor and there must be the audience. And apparently, this sort of thing has gone on since history was first recorded.
The ancient Greeks, the founders of Western rationalism, were particularly addicted to this sort of irrationalism. No Greek would

50   THE SACRED MUSHROOM

embark on any venture of portent unless he first consulted the Pythia, or oracle, or medium, as we say nowadays. The belief was certainly implicit in those times that the gods chose to communicate through the medium of an oracle in order to enlighten man's role in the world of the living and thereby reflect the will of the gods, or destiny.

In modern times this heroic belief has become more the reflection of uncertainty as the drama of the oracle is enacted around the dead and departed. A strong belief in mediumship has been a part of the prevailing spirit of several ages, but in our times it has shrunk to a clandestine practice. And its devotees, strangely enough, are widely diffused throughout all levels of society. What one usually sees in this practice is not too savory, and its low repute is therefore understandable. But occasionally the practiced eye may detect a mediumistic performance that gives one pause. Such was the case as I observed Harry and reflected on what I had seen.

Here was an unvarnished, almost primitive, exhibition of human expression and emotion. It certainly was not a calculated performance that would ensnare one emotionally into an escape from the cares of life, nor offer any panacea to a thrill seeker. The statements were too fragmentary to have any immediate meaning or personal message. It is true that Harry, as Ra Ho Tep, emphasized over and over again that Alice had once been a personage called Antinea in an Egyptian lifetime. But Alice had no remembrance of such a lifetime, and she was much too rational to delude herself into so thinking. In fact, she never did attain any such recollections. The messages in English, or in written or spoken Egyptian, were too enigmatic to have any immediate personal meaning to her.

chapter IV

At half past two in the afternoon of December 13, 1954, a great exhaustion came over me after having been awake more or less continuously for three days. I now had a chance to take a brief nap. Since I was medical officer of the day on a weekend, I knew I would not sleep for long because the phone might ring at any moment, and I would be informed by the Charge of Quarters that I had to attend an accident case or a sick person.

It was the custom of the medical officers on the post, when serving as M.O.D. on a weekend, to stay at home and receive telephone calls there. I was in my apartment which was on the post, and therefore it was possible to take a nap. The phone was at my bedside, and it would only take me about two minutes to arrive at the dispensary if there was an emergency. So I lay down on my bed, after having locked the door, and stretched out fully clothed. It couldn't have been more than a moment or two before I fell asleep.

My next conscious experience was abruptly startling. I found myself floating near the ceiling of the bedroom and looking down on my sleeping body. My surprise was stupendous. When I say that I was floating near the ceiling and looking down at my body with a sense of motion in my limbs, I mean that it was the real I, the one who thought and acted. The body on the bed below me was an impersonal thing. It could have been any sleeping body. It didn't seem to belong to me, and I certainly had very little interest in it. As I looked down at it, I thought rapidly. Could this be some sort of a unique dream experience? Was I really myself as I floated up near the ceiling looking down? I remember distinctly moving my hands over

51

52 THE SACRED MUSHROOM

my "body" and it seemed to be real; the body on the bed *didn't* seem real.

Having spent several months now thinking and studying about the problem of leaving the body I realized that I could be imagining all this. But I also had a unique opportunity to check on my imagination. I felt as though I was coolly planning an experiment in the laboratory. The big question that came to me was how could I prove to myself that this was not a dream? I felt that at any moment I might awaken and the opportunity would be gone. It occurred to me that I should try to "go" to some place, that is, to find out if I could actually observe some distant point. Then perhaps I might observe something which would become evidential, or I might even be observed. Then, I reasoned, when I awakened I might have some objective check on whether I had actually been out of my body or whether I was having an unusual kind of dream.

One of the distinct thoughts which I had at this moment was the feeling that I didn't care if I ever saw that body in the bed again. I had the strong feeling that it wasn't really me; the real *I* was up here near the ceiling looking down on something that could have been a piece of clothing which I had once worn. It flashed through my mind that I must try to go to someone who was definitely sensitive to things that ordinary beings are not sensitive to. I decided to try to reach a well-known sensitive, Mrs. Garrett, with whom I had worked over a period of years in the laboratory. I had confidence that this woman was a genuine sensitive, and if I was at all to be seen it would be by a person of her ability. No sooner had I had this thought than I found myself moving at a great speed through what I can only describe as a gray-blue enveloping atmosphere. I had absolutely no idea of how I left the room. At one moment I was thinking about going to see Mrs. Garrett, and in the next moment I was on my way.

My next experience was equally startling. I found myself in Mrs. Garrett's apartment in New York. I had been to this apartment many times before and therefore I realized I could now be remembering it. I quickly reflected that, as I had traveled from my room in Edgewood, Maryland, to New York I had not observed anything which could be called a landmark on my journey. I was not aware of trees, of skies, of water, of land; in short, it was almost as though I had left the world which I knew when I left my room, entered a world completely unknown to me, and now, as I entered Mrs. Garrett's apartment, had re-entered the familiar world that I knew.

I saw Mrs. Garrett, sitting in a chair, talking to two people. She looked as real as she did when I last saw her. Strangely enough, I could not hear what was being said, although I could see the movement of lips, and felt quite certain that a conversation was in progress. I could not understand why I could visualize so clearly but could not hear anything. The absence of sound intrigued me, but I was powerless to do anything about it.

As I floated around over the three people in the room I remember feeling completely confident that my presence would be noticed by Mrs. Garrett; but she and her companions were completely unaware of my presence. I came closer to her and even waved my hand across her face; still there were no signs of recognition; she went right on talking. The realization came to me that she was not aware of my presence. I even began to feel that my experiment was a failure and I would never be able to know whether I had been dreaming, or whether I had been out of my body on an unknown journey.

Like all human beings, I had experienced dreams in a thousand and one different forms, and if this was a dream it certainly was the most unique dream I had ever known. I felt I had to get an answer as to whether this was a dream. A sense of urgency seized me and compelled me to pursue the experiment.

The only other person that I knew who might be sensitive to my presence would be Alice Bouverie. I had no idea where she was—it simply came to me that I must try to reach her. No sooner had this urgent thought crossed my mind than once again I found myself moving with a sense of speed through a gray-blue misty atmosphere as before. I had no idea how I had left Mrs. Garrett's apartment. It seemed as though the thought was the father of the action. I remember thinking about this in full flight and puzzling over how I had escaped from the three-dimensional confines of a room, into a dimension which even today I cannot understand. Again I arrived in another room with no sense of the mode of entry, and, in this case, without any knowledge of where I had arrived. The room in which I found myself was completely strange; I had never been here before. But there I saw my friend Alice standing in the corner of a dining room with two people, smoking a cigarette and apparently having a conversation. The room was a stately dining room, quite large, high ceilinged, and with a rather unique golden wall covering.

I floated across the room with a rather ridiculous type of swimming motion in order to attract the attention of Alice. I floated

54   THE SACRED MUSHROOM

almost directly over her, waved my hand, and tried to attract her attention. But alas, she took no notice of me. Strangely enough, I recall I made no attempt to reach out and touch her; I did not try to speak, but repeatedly waved my hands, all with no response. A crushing realization swept over me; I realized I could not be seen by others. My experiment was a failure in that it never would have a conclusion. When I woke up, if I ever did, I would have only the recollection of a strange sort of dream. And this was hardly satisfying to one who had devoted years to experimentation and was keenly aware of the necessity of getting objective evidence.

However, rather than admit failure I decided to look around this room carefully and try to remember something that was unique so that I could report on it later and use it as a signature of identification of having been to this place, if it indeed did belong to any place. My sense of vision, which was quite keen in this state, focused on a rather interesting wall covering. I observed that it was made of a golden brocade silk. I remember fixing its design in my mind. This, I thought, should be an unusual and distinguishing mark of this room, since it was rather uncommon. No sooner had I fixed the appearance of the wall covering in my mind than I was startled by a loud noise, and I knew with a desperate urgency that I had to get back to my "body" in the bedroom in Maryland. I have no idea of how I left this room. All I know is that once again I moved through a gray-blue enveloping atmosphere with a terrific sense of speed. I distinctly remember sliding into my body on the bed with a motion which I imagine to be very much like that of a fluid which is suddenly drawn up into a bottle by a vacuum pressure. Immediately thereafter I awakened, still hearing the loud noise which had startled me. I realized that it was the sound of my bedroom door after it had been struck, and I heard the footsteps of my daughter disappearing down the hall. I now realized that she had been at my door and had banged it loudly, perhaps once or twice, and then dashed down the hall.

I calmly lay on the bed looking up at the ceiling. Now I was once again my sodden heavy self. In looking up at the ceiling I could not see any likeness of myself up there. All of me was on the bed. I felt as though I had gently awakened from a deep sleep, and lay there trying to recall what had happened. I realized that every sharp detail of this experience was clear in my mind and that I had lost nothing by the rude event which sent me flying back to my "body." In the

THE SACRED MUSHROOM   55

quiet of my room an uncertainty crept into my feelings. While the memory was vivid, this did not give me any sense of certitude that I had indeed been out of my body. There are many dreams that have a vividness which leads one to believe that it is equal to reality. It is known that many primitive people when they have a dream actually believe that they are in the place of which they dream, and that the people and events that they observe are real. Therefore, I knew from observations of myself, and observations of patients, that a dream referring to objective reality can be a great delusion.

I did within the next few weeks see Mrs. Garrett, and Mrs. Bouverie. I casually asked them what they had been doing on the day on which I had had my unusual experience, and neither of them said or hinted at anything that would lead me to believe that they had been aware of it. However, both had been in the rooms in which I had "observed" them, but this was not of importance as evidence. Mrs. Bouverie thought that the dining room wall covering should be checked. Her mother informed her that about forty years ago the dining room had been covered with brocade, but no one was sure of the color of the fabric. This was the only suggestion I had that my dream could have corresponded to reality. But the dining room walls, which were white now, and had been decorated in brocade forty years ago, offered too much of a time disjunction to consider seriously as evidence.

All in all I felt that my experiment had been a failure. But at the same time I had gained what for me was a unique insight into a human experience which up to now had been only an academic description. Whether I had had a dream, or whether I had been out of my body, I cannot say, and perhaps never will know. But that event truly imprinted upon my mind the conviction that getting "out of the body" and moving around with an ability unknown in everyday life was distinctly within the stock of human experience, whether imaginary or real, and as such probably formed the basis of the traditions and legends found in the literature.

I looked over the references that Harry had made in the trance state to this business of leaving the body. I looked back at all the reading I had done on this subject. I remembered the experience of the two patients who had told me that under nitrous oxide anesthesia they had had the experience of being out of the body. During the next few years I came across several people who confidentially described such experiences to me. All the people that I know who claim

56   THE SACRED MUSHROOM

to have had this experience are eminently rational and sane individuals who carry more than normal responsibilities in life. Some of them are of very high professional standing. Certainly none of these cases which have come to my attention concern individuals who are abnormal, psychologically speaking. I never had this experience before December 13, 1954, and as of the day of this writing I have never had it again.

As I thought about this experience in the weeks that followed, the psychological meaning of it forced itself over and over again into my mind. And gradually there grew on me the conviction that this legend of getting out of the body could well be nothing more than a dream; but if not, it could indeed be one of the rarest and least known of human experiences. I plunged into a renewed study of this phenomenon from a historical point of view in order to find out what had been said about it. These investigations are much too long to burden the reader with, but I can say that from the beginning of recorded history, in every one of the great ancient classical cultures, this experience has been recounted again and again. Mankind has yet to evaluate fully this strange sense of escape from the limitations of time and space.

Although I am convinced that my own experience was only a dream, I cannot discount the evidence cited in the literature of individuals who were seen in places distant from their physical bodies, and who collected verifiable intelligence while in this state.

While I was learning the lessons of this experience, I received another one of those excited phone calls from Alice in New York in the early part of February. As I recall the conversation it went something like this:

"Andrija, I have some hopeful news to tell you. You know I have been doing a great deal of study and research on this mushroom business over in the New York Public Library. Well, just yesterday I was going through some of the books in the Arents Tobacco Room collection, and was talking to the young man in charge of the room. I asked his help in finding some books on mushrooms and other drugs. He seemed rather surprised that I was interested in mushrooms, because he said as far as he knew there was only one other person in New York City who had an interest in mushrooms who also came to this room. I pricked up my ears at this and asked who the person was. He said it was a businessman who was writing a book on mushrooms and who had been working on it for a number of

THE SACRED MUSHROOM 57

years. Would I like to meet him? I said yes, can you arrange for us to get together? He said it was very easily done. I can just call him at his office. So this nice young man put in a call to this gentleman whose name is Mr. Wasson, and I have arranged to meet him. Can you be in New York next week so that we can meet him together?"

"Yes," I answered, "that sounds promising. I do think I could be free next weekend, and if you can make the arrangements I shall certainly be happy to meet this gentleman."

On February 12, 1955, I did meet Mr. R. Gordon Wasson. Alice had invited him to her house for cocktails. I distinctly remember Mr. Wasson's presence that evening as he walked into the room, a vigorous, not-too-tall gentleman in his fifties, and quite distinguished. He seemed a little hesitant about talking to strangers about the nature of his interest in mushrooms. I could not tell whether he was looking upon us as potential rivals or as newly found partners in this lore which he pursued. But my question was soon dispelled, because as soon as he found out that we were interested in the cultural, or ethnic, side of mushrooms he beamed and brightened and began to speak quite freely.

I soon found that I was dealing with quite an expert. I didn't want to leave any false impression with him, so I hastily informed him that my interest in mushrooms was rather recent and that I myself knew very little about the subject. I told him that I was particularly interested in the possible ritual and religious connotations of the mushroom. I told him briefly about the Harry Stone episode as one of the reasons I had become interested in the subject. He told us that he, too, was interested in the ritual use of the mushroom. However, before he went into this subject he explained how he had become interested in mushrooms.

He told us the story of how his wife, who is Russian, had started to write a book about twenty-five years ago on Russian cooking. In reading over the manuscript of the book he noticed some references to mushrooms being used in cooking and suggested to his wife that this point should be clarified by a footnote. His wife, like all good wives, put him to work on preparing this footnote. He soon became engrossed in the footnote concerning a mushroom, and now after twenty-five years that footnote had grown into two large volumes.

However, his most exciting tale had not yet been told. He informed us that in August of 1953 he had journeyed to Mexico because he had heard tales and rumors of a ritualistic mushroom cult which

58   THE SACRED MUSHROOM

supposedly existed there. He had known from his historical researches, particularly the records of the early Spaniards in Mexico, that mushrooms were known to be used for ritualistic and inebriating purposes. But no one knew whether this was historical fancy or whether there was a factual basis for such descriptions.

Mr. Wasson told us that on Saturday, August 15, 1953, in Huautla, Mexico, he had obtained the first evidence that a sacred-mushroom cult was in existence and still practiced in this remote part of Mexico. This had existed long before the Spaniards arrived there; the Spaniards to some extent were aware of it; and the Church had done everything to suppress and stamp it out. The cult practice had gone on apparently for hundreds if not thousands of years in Middle America but had been kept from the knowledge of the Western world and the white man until Mr. Wasson had rediscovered it.

The ceremony in Mexico was conducted by a *curandero*, a sort of unofficial priest, of the mushroom. The *curandero* had to be ceremonially pure in order to undertake the rite, and the preparation usually consumed four days prior to the actual rite. The *curandero* apparently ate the mushrooms during the rite, and the ritual was held for special purposes only. The purpose usually was to help somebody who was sick, to give advice to those in trouble, and to find lost or stolen objects.

When the *curandero* had consumed the mushrooms he then came under the power of the god or gods, and it was no longer he who spoke, but it was the mushroom "himself" who spoke. The mushroom is held in great reverence, and anything that is uttered while under its influence is believed implicitly to be the utterance of the mushroom itself.

I shall not burden the reader with all the elaborate details of this amazing practice concerned with the sacred mushroom which goes on in Mexico, because Mr. and Mrs. Wasson have published their findings in great detail in a magnificent two-volume work entitled *Mushrooms, Russia and History*, published in 1957. The important point about Mr. Wasson's information at this time was a story which he told in connection with his experience with the *curandero* on August 15, 1953, in Huautla.

Mr. Wasson presented this story and its implications rather lightly, but the conclusion was inescapable. The *curandero* did have divinatory powers. I myself was not too impressed by this fact of extrasensory perception. I had actually witnessed the same kind of evidence by great sensitives, who had only their talent to rely upon and did not have the help of the sacred mushroom. Therefore I was not inclined to associate the use of the sacred mushroom and the revelations which proceeded therefrom as being necessarily bound together. I knew from my own observations that one could have extraordinary ability to gather intelligence by means of extrasensory perception, provided one had the appropriate talent. Whether such a talent was actually magnified by use of a mushroom Mr. Wasson had no way of knowing.

This first meeting was all too brief with Mr. Wasson because he had to leave about an hour after he arrived. It was with reluctance that I saw him depart, because he had brought information to me which was extremely important in the light of the material that Alice Bouverie had been recording from the trance utterances of Harry Stone. Mr. Wasson assured us that to the best of his knowledge, which had been world-wide and covered many years, the Mexican sacred-mushroom ritual had not with certainty been known to exist before his discovery. It is true that there were some scattered references to its existence in obscure journals or ancient manuscripts, but no one had ever proven that there was substantial fact behind the legend.

Now, if this were true, then surely Harry Stone could not have come by this information. It was not within his scope or his training to dig up such obscure facts and then try to pass them on as something vitally important and real, as had been done in the trance-state utterances. I was confronted for the first time with an objective proof that there was something to the sacred-mushroom phenomenon which until now had come to my knowledge only through the trance utterances of Harry. At last I had come upon some solid ground in my quest for the sacred mushroom. Right here in the Western Hemisphere, a short airplane ride from the United States, there was a living sacred-mushroom cult which had been observed by a competent witness who had obtained firsthand evidence of the divinatory powers associated with its use.

This revelation by Mr. Wasson set my mind and convictions into motion. I felt that the Harry Stone trances, with their Ra Ho Tep personality and the knowledge delivered therefrom about the sacred mushroom, demanded serious investigation. I made up my mind to try to secure the services of Harry Stone as a resident sensitive in the laboratory in Maine. I now determined to return to Maine as

60 THE SACRED MUSHROOM

soon as my tour of Army duty was ended, which was April 1, 1955, just a few months away. I spoke to my good friend the commanding general and informed him that I was no longer interested in the Army proposal made earlier. I felt that I could pursue my studies with greater ease in Maine than I could within the restrictions of military life. I knew that I could never study a strange and bizarre phenomenon like the Ra Ho Tep personality of Harry Stone within the Army. Such studies were incompatible with the demand for conformity imposed upon government personnel.

I did feel that it was my duty to pass on to the military the knowledge that Mr. Wasson had given me of the mushroom used in Middle America to enhance or fortify telepathic powers. I asked Mr. Wasson's permission to do this, which he granted. I felt that this would satisfactorily close my accounts with my many friends in the Army who had expressed an interest in this subject. What the military did with this information I do not know.

I contacted Harry Stone and offered him a job in Maine. I explained to him that there wasn't much money involved, perhaps fifty dollars a week at the most, for which he would have to work as a subject in the laboratory at least an hour or two a day, and in addition do other work around the laboratory for three additional hours a day. This did not seem too burdensome to him after his strenuous life in a New York factory, and he agreed to do this if I could furnish him some space wherein to carry on his sculpting. Fortunately the laboratory was so arranged that this was possible. Harry agreed to come to Maine and first look over the situation before he made a final decision.

chapter V

HAVING finally settled my relations with the military authorities in regard to whether or not I would continue research in the Army, I began to look longingly toward Maine. Jinny and the children, too, were beginning to get very anxious about leaving a hot southern climate and getting back to the breezy sea-swept coast of Maine. April 1, 1955, came on much too slowly. At last the day arrived when I signed the last separation paper. The furniture was picked up in a huge van by the courtesy of Uncle Sam, and we joyfully piled into the car, including our pet, a black cat whom the children had affectionately named Darkie Motor Boat because he purred so loudly. The trip to Maine took about a week with stops all along the way to visit friends and to enjoy once again the sweet fresh air of civilian life.

In some ways the return to Maine was a great disappointment. In my absence the fortunes of the laboratory had slowly gone downhill. I had started the laboratory in 1948 in a barn in the woods which a grateful patient had loaned me, and by 1952 had built it up to where we had a dozen staff members and a healthy income. After my departure for the Army the research program slowly lost its momentum and the staff members gradually drifted off. Six months before my return to Maine the laboratory had been closed in order to conserve funds. In looking over the account books and the plant, I found that I had a huge job in front of me in order to get the lab operating again.

The plant and equipment had become run-down. Rather than open up the main building, I set up a small office in the barn, which was also used as a laboratory. By working day and night during April

62   THE SACRED MUSHROOM

and May the trustees and I were able to raise enough money to overcome the deficit and have a small surplus which enabled me to hire two staff members, and to give Harry Stone some assurance that I had a position ready for him.

Harry was not quite sure that he wanted to leave the artist's life in New York for a wilderness life in Maine. So on May 14, 1955, he and Betty arrived in Maine just to take a look around to see if they could accept a life as research subjects. The day that they arrived was bright and sunny. They fell in love with Maine, and I knew that they would co-operate in the experiments.

In the meantime, I had made arrangements to bring to the laboratory for a few months another subject, a Mr. Gallow, whom I had observed irregularly during the past two years while I was in the Army. This person fascinated me greatly because he had a very small talent for telepathy but had an enormous talent as a confidence man. He took a great delight in trying to outmaneuver me in tests which were set up in order to study the alleged extrasensory perception which he claimed to have. I had never studied a case where there was such compulsion to be a great sensitive and where all possible means were used to establish this reputation. In order to study him, I, of course, could not yet tell him or anybody else what a problem he was. This would spoil the entire purpose of my study, which was to try to gain some experience in understanding the abnormal side of human behavior as it showed up in pretensions at being a sensitive.

This kind of work was very risky in that my colleagues, and perhaps other subjects, would take exception to such a person when they discovered his shady side. But I took him on as a calculated risk, because it is not easy to get such a subject into a laboratory.

When Harry arrived on the first of June, I settled down to a rather formal program to test various potential abilities he might have as a sensitive. I began with tests to see if he had clairvoyance. Clairvoyance can be defined as the ability of a human being to get intelligence from printed material or other inanimate material. An example of pure clairvoyance follows: a machine prints up some sort of a message by automatic means on a piece of paper and this message is then placed in a closed room. The sensitive is placed at some distance from the printed message, which no human eye has seen. He certainly must be out of the room in which the message has been stored. The sensitive then tries to visualize this printed page clearly in his mind as though he were actually reading it, and then, if he has any ability

THE SACRED MUSHROOM   63

he reads some part of that message. If the sensitive can do this, it is considered pure clairvoyance.

This is a very difficult test situation to achieve. Therefore I compromised in my requirements for a pure clairvoyance test and did the second best thing. I set up a group of ten wood blocks which were two by two inches in size. This set of ten blocks came in duplicate, thus making twenty blocks in all. On one set of ten blocks I asked a colleague to paste ten different photographs, each photograph being distinct; on the other set of ten blocks I asked him to paste a duplicate set of photographs, so that there would be two pictures of each kind in the test.[1]

I asked my colleague not to tell me what pictures he had selected, and to give them to me in a sealed box so that I could not inadvertently look at them. I then got another person to take this finished test and arrange the pictures so that they were lined up in two rows, that is, one set of ten distinct pictures in each row. He was to arrange them in such a way that the pictures opposite each other did not match; in other words, each row represented a random arrangement of the ten pictures.

Now it was the problem of the sensitive, Harry in this case, to take the duplicate set of pictures in the dark and try to match pairs correctly. He could not see them, of course, and they were so arranged that by feeling them one could not tell the difference between one picture and the next. He was to touch one picture in the first row and then attempt to find the picture in the other row that was identical. The idea of the test was to see if he could match the two sets of pictures correctly.

This is a very difficult test to perform successfully. By chance alone, any person, by manipulating the box of pictures, will get zero or one correct match out of ten. If the test is repeated five times, and this means fifty separate matchings of pictures, the chance score for such a series would be five hits out of fifty trials. If the individual gets two or more correct matches each time he attempts to do the set of ten (that is, two correct out of ten trial matches) and maintains this performance for five consecutive runs, he has a score which exceeds chance expectation. His score then becomes what is known among parapsychologists as "significant" statistically. This would be considered acceptable evidence that the person has clairvoyance.

[1] See Appendix 1.

64   THE SACRED MUSHROOM

I ran this test for ten days with Harry. The conditions were such that a colleague of mine delivered the test, which is called the matching abacus test, or "MAT," sealed in a box, so that nobody could see what the arrangement was of the two sets of pictures. Harry and I went into a small room which was especially designed to be absolutely light tight. We sat in this room for about ten minutes in order to get adjusted psychologically to each other and to the atmosphere. Then in the dark I opened up the sealed MAT test and handed it to Harry. When the test was finished the box was sealed in the dark. The blocks were so fixed in the box that they could not move once the lid had been closed, in order that the matching arrangement which Harry had made would not be disturbed. Then the door of the small room was opened, and I handed the sealed MAT test to my colleague, who in turn took it to another room and scored it.

This test was repeated thirty times during a period of ten days, and no one but my colleague knew what the scores were until the entire series was completed. When the series was done my colleague informed me that Harry had obtained what is known as a chance score on this series of tests. In other words, if a monkey had done the same series of tests and arranged the pictures as his fancy suited him, he would have come out with the same score. My conclusion was that, as far as this test illuminated the problem, Harry did not have clairvoyance. Studies over the next two years in other forms of the same test confirmed this opinion. Harry was never able to demonstrate clairvoyance under laboratory conditions.

My next series of tests with Harry was to see whether he had any ability as a psychometrist. A psychometrist is a person who is alleged to be able to hold in his hand an object like a photograph, a handwritten letter, or some object which rather intimately belongs to a person; he is then supposed to be able to get verifiable intelligence about that person merely by handling the object. Since Harry had been reputed to be able to describe what was on a picture when somebody held it in his hand across the room, I thought I had just as well find out if he really had this ability. The ability to tell a person what is on a picture when the owner is holding it in his hand (and thus knows what is on the picture) would be called telepathy, or mind reading. Psychometry is a more rare phenomenon, and there are not many people who can actually demonstrate this talent. This series of tests with Harry brought me to the conclusion that Harry was not a psychometrist, but in the course of the tests I found that he showed

THE SACRED MUSHROOM   65

what could be classified as telepathy. Therefore I turned my attention to testing him for a very simple form of telepathy.

The form of this test was as follows: Harry was asked to leave the room; a monitor left with him, and he securely blindfolded Harry before he re-entered the room. While Harry was gone somebody present was asked to hide an object somewhere in the room. The object could be as small as a dime and it could be hidden anywhere. In doing this test with Harry many times it was found that as long as somebody in that room knew where the object was hidden Harry could always find it. If, however, special precautions were taken so that nobody in the room knew where the object was hidden, Harry could not find it. Now the first instance would be a demonstration of simple telepathy, and the latter case would fall under the heading of complex telepathy, although this is a rather loose definition of this ability. With this finding, namely, that he did have talent for a rather simple form of telepathy, the studies shifted to a formalized test for telepathy.

The test for telepathy was the technique already described for pure clairvoyance, the MAT, with some modifications. The test was now so arranged that the ten pictures were strung out in two parallel rows on a rectangular table about the size of a coffee table, and covering these pictures was a sort of a chicken-coop housing. The chicken-coop house was so arranged that Harry, sitting on one side with his hands under the roof, could touch the pictures but could not see them. The person sitting across the table from Harry, by looking under the roof of the house, could clearly see where the pictures lay and their relationship to Harry's hands as he sought to match the pictures in the two opposing and parallel rows. Special precautions were also introduced by securely blindfolding Harry so that he couldn't see the person sitting opposite him, and other arrangements were made so that he could not be aware of any unconscious auditory clues that the person across the table from him might give in helping to select the correct matching picture.

In repeating this telepathy test hundreds of times with Harry I found that he did indeed show clear-cut evidence for telepathic ability. In doing the test under normal room conditions, where the possibilities of fraud, unconscious clues, or hypersensory perception on Harry's part were eliminated, he was able to average about twelve correct matches out of each set of fifty trials. And this is considered proof that he did have a gift for telepathy.

66   THE SACRED MUSHROOM

Now while these tests were going on with Harry, similar telepathy tests were going on with Mr. Gallow. However, I made a point of doing the work with Mr. Gallow privately and not allowing Harry to know what the results were. In doing the same test for telepathy as I have described for Harry, Mr. Gallow was able occasionally to achieve under rigid test conditions scores of eleven correct matches out of fifty. And this is borderline evidence that he did occasionally have telepathy. However, Mr. Gallow was never satisfied with such marginal performance. He wanted to be a great star and a great sensitive, and so he contrived endless ways in which to cheat on any test which I presented to him. For example, in the case of the telepathy MAT test he would contrive ways of marking the individual pictures so that he could feel which pictures matched. As soon as I detected this, I, of course, changed the pictures. Then he would try new tactics. He would put pin holes in his blindfold, or distort it in such a way that he could peek alongside his nose. Or he would fish for verbal clues continually while doing the test. And each time he took a new gambit I would tighten up the test conditions in an unobtrusive way. His daring was amazing, and he taught me a great deal about the fine art of gamesmanship in ESP testing.

Occasionally I would allow him more latitude in order to see what particular device he had worked up anew. And under these conditions his fraudulent score would run up as high as I allowed it to go. And he delighted in getting such high scores. These scores, of course, were not obtained by extrasensory perception, but by deception. Mr. Gallow, when he achieved such a score, could not help but go and brag to Harry and others as to what a great sensitive he was. This bragging had a very adverse effect on Harry. I could see that he felt he was in an unwholesome atmosphere, because his intuition told him that Mr. Gallow was not the great sensitive he advertised himself to be. And over a period of weeks I could see that this was developing into a justifiable resentment on Harry's part.

One morning I walked into Harry's studio to see how he was getting along. I found him in a very black mood, packing his belongings. This was the first time that I had ever seen him so angry. Realizing that here was a major problem, I approached him quite delicately and inquired as to the reason for packing.

Without looking at me, he bluntly stated that he was leaving the laboratory. When I asked him if it was anything that I had done to bring this about he turned to me and replied that the answer was

THE SACRED MUSHROOM 67

both yes and no. This left me more puzzled than ever and I persisted in my query.

He finally announced quite bitterly that he could not stand the pretensions of Mr. Gallow. He felt that I was being taken in by Mr. Gallow and that he was being made a fool of in the laboratory tests.

When I inquired as to the reasons for this opinion, he admitted he didn't have any proof, but that he was certain that Mr. Gallow did not possess the ESP talent that he claimed. In any case, Mr. Gallow made him feel quite shabby about his role in the laboratory because his sincerity was being weakened by Gallow's opportunism and insincerity.

Harry made it quite clear that he was not fooling, he was going to leave. I had no other choice but to explain to Harry that he was quite right in his opinion, and that I was fully aware of Mr. Gallow's opportunism and insincerity in his general character and in the laboratory work. But I made it quite clear to Harry that my job was to study the personality problems of sensitives, as well as genuine extrasensory perception. Mr. Gallow was one of my pathological cases and it was important that I study him until he had exhausted all of his art of illusionism. When I had found out all I could about his abnormal motivation and his stock of tricks, my study would be completed, and Mr. Gallow's stay would be terminated. I begged Harry to be patient with me, as well as with Mr. Gallow, and stay on at the laboratory. I pointed out to him that he should not allow his own sincerity and integrity to be affected by his association with this man.

Harry breathed a sigh of relief, and felt much better in knowing that I was not being deceived by Mr. Gallow, nor was I in any way associating his approach and his work with that of Mr. Gallow. He agreed to stay on.

I respected Harry's forthright position in making it clear to me that he did not want to be a part of anything that was not completely honest and sincere. He returned to his work with a much more objective attitude about Mr. Gallow's great pretensions.

Harry had not done anything really remarkable up to now as a sensitive in the laboratory. His Ra Ho Tep manifestation had ceased as of December 9, 1954, and since that time he had not either gone into a trance, or spoken or written any Egyptian. I had discussed this situation a number of times with him and asked him what he thought about it. He said that he just didn't know, because when the thing had happened, he had had absolutely no control over it.

68 THE SACRED MUSHROOM

And now that it wasn't happening, he had no control over it either. In fact, he said, he was quite glad that the thing had disappeared because he was beginning to worry about his mental stability, especially when told afterwards what had happened when he had suddenly slipped off into trance. I could sympathize with his feelings because I certainly would not want to be the victim of sudden trances or amnesia for an event. Harry's real happiness in being at the laboratory was the opportunity to devote a great deal of time to sculpting. He was grateful to be out of that mannequin factory in New York.

One day Mr. Gallow had been kidding Harry by suggesting to him that he had better slip into trance once in a while and "Make with that Egyptian stuff," or otherwise no one would be interested in him. Harry had quite a heated discussion about this, and told him in no uncertain terms that it was not an act and he was very happy not to be bothered with it. This incident was reported to me by Betty, and I had no reason to doubt it, knowing Mr. Gallow's intriguing attitude. I developed quite a respect for Harry's integrity after observing him for a month.

Since Harry had arrived in Maine I maintained an attitude of not encouraging his Ra Ho Tep manifestation. This was a continuation of the policy which I had suggested to Alice and which she had followed. My work with Harry now was solely along the line of laboratory testing for the presence of clairvoyance and telepathy. On June 14, 1955, while we were doing a MAT test Harry abruptly went into trance for the first time in Maine. This time he said nothing but indicated that he wanted to have a paper and pencil which I promptly handed to him. At the moment he was fully blindfolded and drew on the paper a picture of a human figure with a solar disk over the head, and over this a pair of horns. This is all he did during this trance session. I interpreted this figure as representing either Tehuti or the god Shu. In this form Tehuti would be represented as the moon god. Shu is one of the most ancient of Egyptian gods and as the god of air he held up the sky.

On June 20, 1955, while I was doing routine laboratory tests with Harry, he again unexpectedly went into trance. During this trance he gave no indication of wanting to write anything, but uttered only two phrases in Egyptian. The rest of the trance performance, which lasted thirty-five minutes, was a long series of dramatic sign-language gestures which were hard to understand and follow. It was my impres-

THE SACRED MUSHROOM 69

sion that I was observing a silent ritual form which was concerned with the use of the sacred mushroom.

However, the two Egyptian phrases that Harry vocalized I can understand and discuss. In the beginning of the silent ceremony Harry made the statement: KA, HOP [P or B], NOU, MI. I translate this phrase as meaning: "The spirit of Tehuti appears," or this phrase could have an equivalent translation as "The spirit of Tehuti looks on." The second word is pronounced HOB with a hard labial stop on the B sound, and HAB is known as one of the common names in ancient Egypt for the god Tehuti.

Later on in the silent ceremony Harry uttered the word "SHU." This phrase was uttered after that section of the ceremony which I interpreted as being the part where the person has just received the sacred mushroom. I associate the word SHU with the air god Shu.

The important thing about this demonstration was that it had little immediate meaning to me, and it certainly was not calculated either to illuminate or to mystify. It was so far beyond anything that one might get remotely interested in that I could only come to the conclusion that there was no ulterior motive behind this demonstration. Harry himself was totally unaware of it afterwards; in fact, he awoke from his trance and went right on doing his test, saying only that he felt drowsy. I said nothing to him about what had happened. He never mentioned the subject to me. So I certainly had no grounds in this first demonstration of his trance condition since December 9, 1954, to suspect he was in any way trying to impress me. On the other hand, the almost total cessation of the trance phenomenon and its cryptic messages were a puzzle to me. All I could do was quietly stand by and wait to see what would happen while going on with routine experiments in extrasensory perception, and try to get to know and understand Harry's history and character.

chapter VI

IN the early part of June I had a meeting with Mr. Wasson in New York. Mr. Wasson described in detail his plans for another exploratory trip to Mexico in the latter part of June. He was most generous in inviting me to come along on this trip to observe at first hand the strange ritual of the sacred mushroom in Mexico. However, I had my hands full organizing the laboratory and getting a research program under way after my two years' absence. So I regretfully declined his invitation and instead suggested that we set up some sort of an extrasensory-perception experiment between Maine and Mexico. I proposed to Mr. Wasson that he should try to direct a *curandero's* attention to the laboratory in Maine in order to find out if he could divinate by extrasensory perception what was going on. In the meantime, we in Maine would set aside certain periods and carefully record situations that we would deliberately set up and that would serve as targets for the alleged ESP of the *curandero* in Mexico. Mr. Wasson agreed to this proposal and said he would try to find such a *curandero*.

Now the circumstances were naturally rather good for such an ESP experiment. Mr. Wasson had never been to Maine and had no idea what the physical plant looked like, and, furthermore, had little idea as to what constituted laboratory procedure in parapsychology. This lack of knowledge on his part would minimize the amount of information that a *curandero* could get from him by direct and simple telepathy. In other words, if the *curandero* was able to get any accurate intelligence about what was going on in Maine he would have to do it either by long-distance telepathy or by clairvoyance. Both

72   THE SACRED MUSHROOM

Mr. Wasson and I tacitly understood that neither one of us should know what the other one was doing. I did not tell Mr. Wasson what plans I had in mind for the period of this test, and his only duty was to send me a letter or a telegram from Mexico once he had found a *curandero* who would perform for him; and thereafter we would perform in Maine on our own schedule. There was to be no pre-arranged time when we were supposed to do our part of the experiment in Maine, or a time at which he was supposed to do his part in Mexico. In fact, Mr. Wasson was going to a part of Mexico where he would be completely out of touch with civilization as far as telephone or other rapid communication facilities were concerned, and this made precise synchronization impossible.

With this arrangement made between us I went back to Maine and planned the kind of experiment that would be done. I knew that with the subjects I had on hand, namely, Harry Stone and Mr. Gallow, we did not have good enough sensitives to be able to hope to get any ESP intelligence about what was going on in Mexico, although we would make a try in this direction. Our main effort was to be concentrated on setting up unique and simple situations which were to be fully monitored as far as the time of the event was concerned, so that if the *curandero* in Mexico did, with the help of his mushroom, obtain any intelligence we could rather precisely correlate the two acts, both as to event and as to the time of occurrence.

With these conditions before us I assembled a group of ten people who would participate in the experiment, the reason for the large number being that the ceremony in Mexico usually started at 9:00 or 10:00 P.M. and ran all night long. This meant that we in Maine would have to conduct our part of the experiment through the long hours of the night. In addition to this we had our normal daytime duties, and since we didn't know whether the experiment would go on for two days or two weeks we had to have some reserve manpower in order to carry the burden of a round-the-clock schedule.

We realized that the *curandero* in Mexico would be a simple unlettered native who would have no intellectual comprehension of the complexities of modern civilization or of elaborate laboratory techniques. Therefore, we decided to set up target situations which were within his grasp and within the range of his everyday experience. To this end we decided that it would be sufficient to do small things, and small ceremonies, at stated intervals during the night,

THE SACRED MUSHROOM · 73

and these would be the events that he would have to describe by ESP. We planned to set up for example, the kind of small house-shrine that every Mexican home has. These usually centered around the picture of a saint or the Madonna and are surrounded with paper flowers and a few burning tapers. Other situations were to be group singing, where we would sing simple folk songs, and in different languages if possible. Another situation which we planned was small-group prayer meetings, and other rites which the *curandero* might be familiar with that went on in a church. We also planned to play simple games such as checkers, Scrabble, cards, and perhaps even the Chinese I Ching.

In order to avoid fatigue we planned to carry on these simple procedures in small shifts of two or three people during the night, each taking a two-hour stint. We had scarcely completed the arrangements for the Maine-to-Mexico experiment when I received a letter from Mr. Wasson on June 29, 1955, saying that he was leaving Mexico City, and was on his way to Huautla. Our purpose in this experiment, if such it can be called, was to find out if the sacred mushroom of the Mexicans was an answer to the problem of extrasensory perception posed by the colonel a year ago. This problem had been in my mind since the Army days and had been magnified and stimulated by the material which Harry had delivered in the trance state. Now we were going to attempt to get objective evidence as to the reality of what I had only academically explored in books, what I had interpreted from the Ra Ho Tep phenomenon, and from the evidence that Mr. Wasson had brought back from Mexico.

On June 30, 1955, the first of the nightly experiments in connection with the Mexican project was begun. On this evening while I was conducting some rather simple procedures with the staff, Harry was standing by watching us when he unexpectedly slipped into trance. One of the staff handed him a pad of paper and a pencil, as Harry indicated that that was what he wanted. Then Harry filled eight pages of paper with hieroglyphs, which is the longest continuous series of Egyptian writing that he had ever done. I present the translation of the remarkable series as follows:

LINE ONE: *This is the name of the place of waters from heaven.*
LINE TWO: *The purified noble waters.*
LINE THREE: *I have made the offering of the noble waters.*
LINE FOUR: *The waters from heaven from the mouth of the King.*

74 · THE SACRED MUSHROOM

LINE FIVE: *My name is the Waters from Heaven, waters in an alabaster vessel.*

LINE SIX: *I have made the offering of the noble water in an alabaster vessel.*

LINE SEVEN: *In my name of Ren Ho, Waters from Heaven.*

LINE EIGHT: *The alabaster dagger.*

LINE NINE: *Cut [or reap] with the noble alabaster dagger.*

LINE TEN: *The flowing water from Ra [drink it].*

LINE ELEVEN: *The noble alabaster dagger.*

LINE TWELVE: *Drink from the place of flowing waters.*

LINE THIRTEEN: *The flowing grains [Incense? Water?].*

Now this passage is very obscure in terms of modern thinking, but when placed in the context of ancient Egypt it does make some sense. For example, the entire tone of this litany is not too unlike the tone of the litanies found in the ancient Pyramid Texts.

"The noble waters from heaven" is the central theme of this litany. We must remember that just as the rite of baptism in the Christian Church is believed to wash away original sin, so the ancients believed that purification by water was an essential preparation for any important religious ceremony. Hence we may assume that this reference to water from heaven means in general some preparatory purification process which is to take place.

We also notice that there is a reference in this litany to an alabaster dagger with which one cuts or reaps something. Now a sickle in ancient Egyptian times was made with flint cutting edges. One could assume that the alabaster dagger might just as well refer to a flint knife and as such be used as a sickle. The question was: What is to be reaped or harvested?

The only connection I could think of was that in western Europe there was, long before the Romans recorded it, the practice of reaping certain "magical" plants with special knives or sickles.[1] For example, the Druids of western Europe believed that a golden plant which grew upon the oak was sacred. This plant, which is believed to be the mistletoe, appeared very rarely in the form desired by the Druid priests. When such a plant was found, the priests made a great ceremony out of harvesting it. They did this, so Caesar records it,

[1] The ancient Syrians had a sacramental dagger, or sickle, which they called the AKHW. Its exact ritual function is not known. *Les Reliques de l'art Syrien,* Pierre Montet. Strasbourg, Fascicule 76, 1937, pp. 35-36.

THE SACRED MUSHROOM   75

with a golden sickle and a white cloth placed on the ground under the plant. One of the priests climbed the tree and cut off the plant so that it fell upon the white cloth. From this practice, known in Europe, and which is known elsewhere, we can infer that some special ceremony may also have been involved in harvesting or reaping the sacred mushroom. It may well have been that an alabaster dagger or flint sickle was used for this ceremonial purpose.

We note in lines twelve and thirteen some association between the place of flowing waters and the flowing grains. Now the "flowing grains" could be drops of water, and this description is used in the hieroglyphs of the Egyptian language. On the other hand, the idea of using "alabaster daggers" for harvesting, in connection with the "flowing grains," might also refer to the mushroom with its well-known grains of warts and spots on the cap.

The place of flowing waters, besides the meaning of purification which we have discussed earlier, may also be a symbol for the ascent to heaven. For example, Shu, as the god who held up the sky, was also the god of the air. As such his vapors ascended from the earth to heaven and, of course, descended in the form of rain. It was necessary, to some extent, to get into the framework of the thinking of ancient Egypt, if this is at all conceivable, in order to understand the peculiar archaic character and thinking of the writings which appeared through the entranced Harry Stone.

On July 1, Harry, in the wide-awake state, did some introspection and free association and insisted to all of us present at the laboratory that he felt there was a great urgency that we pour water all over ourselves. He did not know why he felt this way but he felt that it should be done. Of course, none of us could accept this suggestion. As a matter of fact, both he and the rest of us were completely unaware at this time of the meaning of the hieroglyphs which Harry had written on the previous day. In other words, Harry was saying something out of his own consciousness which he had produced through the different entranced state of consciousness in the ancient Egyptian hieroglyphic writing the day before.

Since Harry had been making such a fuss about pouring water over somebody, I decided to take him aside and get him calmed down. I led him into a small room and asked him to lean back in an easy chair, close his eyes, freely associate his thoughts, and tell me what he saw. The first thing he saw was a water pitcher. He just couldn't get that idea of pouring water over somebody out of his mind. But now

76   THE SACRED MUSHROOM

he saw the water pitcher in front of the door to eternity. He again saw the water pitcher in his mind and insisted that he must pour water all over himself. I interpreted this again as being a ceremonial purification, but for what purpose I did not know.

I interrogated Harry about the meaning of the door that he saw and the water. While I asked this question Harry seemed to slip into a deep trance suddenly, and he blurted out a single phrase which sounded to me as follows: ASH NU AH. I watched him for a few moments and saw that he had again come out of the trance. I did not pursue this strange phrase that he had uttered, but continued my interrogation about the door and the water pitcher. Harry apparently wasn't aware that he had briefly been in trance. He went on talking about the door, and said that now he saw a man standing there with a mask on his head. The mask was that of the head of a dog with a very long nose. I asked him if he had any idea what this dog meant. He said that he had no idea. His only comment was, "It's going to be a sacred happening." This ended the interrogation and the interview.

The meaning of the phrase ASH NU AH was quite a mystery to me for several years. I finally discovered that the reason it was difficult to translate was that it appeared only once in the entire body of Egyptian literature, and this in the Pyramid Texts. It seems that the word ASH NU AH is a compound word. The first part ASH refers to the name of a dog that pulls the sun boat, the sun boat, of course, being the vessel by which the deceased ascended to heaven as he went over the "water." We also know that the Egyptians believed that the deceased ascended to heaven by a staircase, by a ladder, and on the wing of either Tehuti or, in some cases, on a cow. This word for a sacred dog, ASH, is a word that is believed to come from the west of Egypt.

The second part, NU AH, is also a word that appears only once in the Pyramid Texts. This again is a name of a dog god who pulled a sun boat. And in the Pyramid Text the deceased is referred to as the "Great NUAH," that is, he is personified as this dog who pulls the sun boat to heaven. In the Pyramid Texts, the deceased is first purified by water, as he begins the boat ascent to heaven in the form of the dog god called the Great NUAH. Thus it was very curious that Harry, in the trance state, should associate two names; one whose origin is to the west of Egypt, and the other whose origin is to the east of Egypt, both meaning the dog who pulls the sun boat.

THE SACRED MUSHROOM 77

And more significant is the fact that both words are extremely rare. In the light of Harry's great insistence on purification by water, it's interesting that in that part of the ceremony in the Pyramid Text where the dog god appears, the deceased is first supposed to purify himself by water before he begins the actual ascent to heaven.

On July 3, 1955, we began the fourth of the Maine-to-Mexico all-night experiments. Three days and nights of staging experiments had left all of us exhausted. We really did not have much energy or enthusiasm for this drawn-out type of work. About ten o'clock on this evening I sat alone with Harry in a room to see what he would do. Neither one of us said a word for about ten minutes, and then I noticed that Harry had quietly slipped into trance once more. He made no gestures, he made no attempt to write, he just quietly sat there. Then he opened his eyes, and I could see that he was still in deep trance. He looked at me with a penetrating gaze and said in Egyptian, and I quote: "NA HA HE HUPE."

The translation of this phrase is: "We are under the care of Hupe." Hupe is one of the names of the great sphinx at Giza near the Great Pyramid of Khufu. The meaning of this phrase is not clear to me even now. A few moments later Harry uttered another Egyptian phrase which I give as follows: "NU AH A HADI." The translation: "The great water enters the tree [the tree as a plant overlaid with gold]." This phrase would seem to indicate that the waters referred to earlier, that is, the noble waters from heaven, were about to enter a tree; he gave the name of this tree in Egyptian as the AH tree [or a plant?]. The particular word that he used for a plant, namely, HADI, can also phonetically stand for a word which means "something which is overlaid or gilded with gold." Not knowing which meaning was referred to, I have included both meanings, namely, the plant and the idea of something covered over with gold.

After this sitting Harry said that he was extremely tired and didn't think that he could go through another all-night session. So he lay down on a couch in the main laboratory where we were carrying on the demonstrations and promptly fell asleep.

Not only was Harry exhausted, but the others on the staff also begged relief because of extreme fatigue. Mr. Gallow fell asleep in his chair. Betty decided to retire. Henry, Graham, and Leo, the staff members, begged off and said they also would retire. That left Alice and me awake in the main laboratory, and Harry and Mr. Gallow asleep beside us. Both Alice and I felt that we must do something

78   THE SACRED MUSHROOM

that the *curandero* could use as a target in case a ceremony was being performed tonight in Mexico. Since we were also exhausted, we decided to play a game just for relief, which would still be a sort of target for the *curandero*.

First we played checkers for about twenty minutes, but this became boring. Alice then suggested that we carry out a mock-hypnosis experiment on either Harry or Mr. Gallow, who were sound asleep, since this would be an interesting target for the *curandero*.

The idea didn't particularly appeal to me because I, too, was very tired. But in the absence of a better idea it seemed to be worth a try. I sat down beside Harry while Alice took up a pad and pencil to take notes. I began to speak to the somnolent Harry, and told him over and over that he was sleeping deeper and deeper and that he was going farther and farther back in time. He went on sleeping but breathed deeper and deeper with marked abdominal breathing. The monotony of my own voice was beginning to make me sleepy, and I snapped my head up to see if Alice had noticed it. As I looked at her I saw that she was asleep but still sitting upright and writing.

I looked at her pad and noticed that she had spelled out a word, which was OST. I asked her what this meant. She mumbled back that she was spelling out a message. This puzzled me greatly.

I looked sharply at her and said, "What are you doing? You're not supposed to be hypnotized."

She sort of muttered as though half asleep and said, "I'm playing a game. I have to spell out these words."

"But," I said, "those words have nothing to do with a game. What does OST mean anyway?" She paid no attention to me and went on writing and spelled out the word TIRIAN. I looked at her more closely and decided that she herself must be in a hypnotized state, so I just pretended that I was playing the game with her.

I asked her, "What does TIRIAN mean?"

She wrote out "My nationality."

I said, "Oh, you are Tirian, and your nationality is represented by that word? But I do not know what the word Tirian means. Could you please spell it again?" Alice then wrote a new word, which was SIRIAN. She kept on writing, and I present her answers in reply to my questions:

A. P. *Oh, this is a nationality? Could this perhaps be Syrian?*
(*No answer.*)

THE SACRED MUSHROOM   79

A. P. *Are you still awake?*
A. B. *No.*
A. P. *Did you ever live before?*
A. B. *Yes. Nineteen hundred.*
A. P. *Where were you born?*
A. B. *Arabia.*
A. P. *Are you interested in the sacred-mushroom ceremony?*
A. B. *Yes.*
A. P. *Did you know this mushroom in your lifetime?*
A. B. *Yes.*
A. P. *Was the mushroom known to your country?*
A. B. *Yes. Syria.*
A. P. *Where were they found?*
A. B. *Amanus Mountains.*
A. P. *What species of mushroom did you find?*
A. B. *Am-amanita muscaria.*
A. P. *What was their color?*
A. B. *Red.*
A. P. *How did you use the mushroom?*
A. B. *Piquer.*
A. P. *What did you dissolve the mushroom in?*
A. B. *Sulphur.*
A. P. *What effect did this have on you?*
A. B. *Trance.*
A. P. *Were you ever successful in getting out of your body during your lifetime?*
A. B. *Yes.*
A. P. *What was the site of injection of the mushroom you prepared?*
A. B. *Tobroquaine.*
A. P. *Is this phrase in English?*
A. B. *Yes.*
A. P. *How many words does it constitute?*
A. B. *Two. Tobro Quaine.*
A. P. *I'm sorry, I do not understand this phrase. Can you get a message from the curandero in Mexico?*
A. B. *No.*
A. P. *Will Mr. Wasson bring back some specimens of the sacred mushrooms?*
A. B. *Yes.*

80 THE SACRED MUSHROOM

A. P. *Will Mr. Wasson see the place where the sacred mushrooms grow?*

A. B. No.

A. P. *Is the mushroom that you know and have used in your lifetime to be found in Maine?*

A. B. Yes.

A. P. *Can it be found locally, near here, in Maine?*

A. B. Yes.

A. P. *Can it be found right here in Glen Cove?*

A. B. Yes.

A. P. *When can it be found?*

A. B. July.

A. P. *Can it be found in the woods?*

A. B. Yes.

A. P. *Can it be found in the fields?*

A. B. No.

A. P. *Can it be found under the oaks?*

A. B. Yes.

A. P. *Is there a performance tonight of the* curandero *in Mexico?*

A. B. Yes.

A. P. *Are you in contact with the situation down there?*

A. B. Yes.

A. P. *Are they using brown or red mushrooms?*

Alice made no response to this question. Instead she wrote out the following:

A. B. *"Will contact Harry at 5:00 A.M."*

A. P. *Is this a message for us?*

A. B. *"A. B. could find mushroom."*

A. P. *Where could she find the mushroom?*

A. B. On the coast.

A. P. *Do you mean right here on this peninsula?*

A. B. Yes.

A. P. *Do you mean on the Glen Cove side of the peninsula?*

A. B. No.

A. P. *Do you mean on the Penobscot Bay side?*

A. B. Yes.

A. P. *Will it be near the road?*

A. B. Yes.

A. P. *Will it be near the north or the south end of the road?*

A. B. South.

THE SACRED MUSHROOM   81

At this point Alice suddenly snapped to and awakened. She looked around and said, "What's been going on?"

I couldn't help but laugh gently, "Alice, don't you know what's been going on?"

Wonderingly, she replied, "No."

"Alice," I said, "do you know that you have been writing out words on your pad while I have been asking you questions, and it almost looked like an automatic writing performance?"

I showed her the notes she had made. She was quite taken aback, and said she hoped she wasn't getting like Harry and coming under the spell of "outside" agencies. I assured her that this probably wasn't the case, that she had just been half asleep and had responded to my attempt at hypnotizing Harry by being hypnotized herself, as so often happens.

"Furthermore," I said, "this message contains a prediction. It said that you could find the mushroom. The real test of what just happened would occur if you ever found the mushroom we are seeking!"

"Yes, that would be quite fantastic," she said. "I don't see how it could ever happen in this world."

At the end of this performance, and it was now 2:20 A.M., I decided to go into the kitchen and make some coffee. Alice and Mr. Gallow joined me. Harry was left asleep. He joined us in the kitchen at about 3:20 A.M. None of us said anything to him about what had happened while he had been asleep. Only Gallow, Alice, and I knew what had happened. We reassembled again in the laboratory about 4:40 A.M., and we sat around sleepily trying to figure out what to do next. I asked Harry if he would join me in a game of Scrabble. He said he would prefer to go to bed, but if we must do something, Scrabble was about the easiest thing he could think of. So we played Scrabble starting at 4:40 A.M., and sharply at 5:00 A.M. Harry went into a trance state and began shuffling the Scrabble blocks around rapidly to spell out the following message which I recorded as fast as he arranged the blocks: ILLOSCHUVIQUTERO TUNAOOTALIIMIA, THREE, TORROESTADENEGRO.

After spelling this out at quite a fast pace, Harry abruptly woke up, looked at me, and said, "What did I do?" "Harry, you have been spelling out words with the Scrabble blocks." He looked at me rather sleepily and said, "Now I know it's time to go to bed. Good night!" This ended the demonstrations for the Maine-to-Mexico experiment on July 4, 1955.

82 THE SACRED MUSHROOM

The letters which Harry had arranged so rapidly meant nothing intelligible. He had probably unconsciously heard us say something about his being contacted at 5:00 A.M., and had responded by shuffling the Scrabble blocks. I thought so little of this entire performance that I laid aside my notes and completely forgot about them during the day, which was the fourth of July, and also on the fifth of July. I didn't think of it again until the sixth of July. On this day we got up very late and were sitting around the kitchen about eleven in the morning discussing the events of the past few days. Alice was much more intrigued by her automatic writing performance, of course, than I was; and she wondered if we hadn't better do something about looking for mushrooms.

Now I must describe the ground layout of the foundation property. The laboratory was located on a peninsula which was a mile long jutting out into Penobscot Bay, which is on the Atlantic coast. The landward side of the peninsula is bounded by a cove called Glen Cove, from which the town derives its name. The outer side of the peninsula fronts Penobscot Bay and the Atlantic coast. The entire peninsula had on its water periphery an old gravel carriage road. And this is the road referred to in my questioning of Alice on that morning of the fourth of July. The answer she had given me was that the mushroom could be found at the south end of this road. This was the edge of our property. If one followed the directions in looking for mushrooms he would have a relatively small area to explore, perhaps fifty yards wide and a hundred yards long. This would conform to the description that Alice had given.

Alice said that she would like to take a walk in the woods to see if there were any mushrooms growing. I directed her to this area of the road at the south end of the property which was about a half mile from the laboratory building. Alice, Harry, and Betty decided to go on this expedition and get a breath of fresh air. I stayed behind and busied myself going over the laboratory procedure of the past few days and checking my notes. The three of them were gone for about two hours.

They all burst into the kitchen in great excitement about one o'clock and proudly showed me a mushroom that they had found near the road at the south end of the property. I must say that the mushroom was of a rare beauty. It was of a golden color, the cap was about six inches across, and the stem was about eight inches long. The cap had sprinkled over it about thirty or forty yellowish-

THE SACRED MUSHROOM 83

white warts. We all looked at each other im amazement and practically asked the same question at once, "Could this be an *Amanita muscaria?*"

I examined the mushroom very carefully and found that it was as fresh as it could be. In my judgment it was no older than twenty-four hours. In other words, it had sprung up within the last two days. I rushed to my mushroom book and carefully compared every detail of this mushroom with the description in the book. The mushroom fitted exactly the description given for *Amanita muscaria,* as did the spores, which I examined under the microscope. This, however, was not sufficient identification. I knew that the true *Amanita muscaria* was supposed to kill flies. So we collected ten live flies in order to see if they would be poisoned by eating the mushroom. The mushroom was mounted on a little base stand and the entire mount was placed within a large bell jar. Into the bell jar were introduced the ten live flies. After three hours of confinement in the bell jar with the mushroom, by actual count we found that six of the flies were dead. Twenty-four hours later eight of the flies were dead, but two still remained alive. There was no question that this mushroom was toxic for flies.

I took a few of the warts, ground them up, mixed them with honey, and placed them near a beehive that we had on the foundation property. I noticed that the bees were attracted in great numbers to the honey bait. After six hours I counted hundreds of dead bees around this mushroom-honey preparation. There was no question that the honey was toxic for bees. The conclusion was slowly forced upon me that we had a genuine specimen of *Amanita muscaria.* This was confirmed for me later by a botanist who examined the specimen and the spores.

Was there any relationship between Harry's hieroglyphic message; the written message that Alice had given on July 4, and the finding of this specimen of *Amanita muscaria* on the sixth of July? Now it is to be noted that in the summer of 1954 I had written to the Boston Mycological Society inquiring if the *Amanita muscaria* could be found in Maine. The reply from the society was that some of their members had occasionally found specimens in New England, but no recent reports had come in of any found in Maine. I checked the literature and found that in one report, published in 1925, a specimen of *Amanita muscaria* had been found in Maine. I, myself, had

84 THE SACRED MUSHROOM

never, during the summer of 1954, or up to this time in the summer of 1955,[2] found any *Amanita muscaria* in Maine.

My general impression was that the *Amanita muscaria* was uncommon in Maine. Could this be one of those rare coincidences where something had been predicted and two days later the event had occurred as advertised? I felt that the relationship between the prediction and the event was probably pure coincidence. But the only way to answer this question was to go out and seek for more specimens of the *Amanita muscaria*. Therefore, we organized a party of ten people and over the next four days spent many hours in combing every square foot of the peninsula on which we lived and which comprised about six hundred acres. No other specimens of *Amanita muscaria* were found on this peninsula. Over the next two weeks the search was extended to include an area of ten square miles around the foundation property in those spots where it appeared likely that mushrooms would be growing. No other mushrooms were to be found in this extensive search. The conclusion was forced upon us that this was not pure coincidence, the finding of the *Amanita muscaria*. While none of us could take Alice's automatic writing seriously, we could not help but puzzle over the relationship between this message and the finding of the mushroom.

The finding of this beautiful golden specimen of *Amanita muscaria* mushroom was a noteworthy milestone on the quest which had begun over a year ago for such a mushroom. A year ago I was introduced to the practice of the sacred mushroom by a very strange message, partly in English and partly in ancient Egyptian, sent to me from New York by Alice. Subsequently I had confirmed for me the fact that the ancient hieroglyphs were indeed genuine Egyptian writing. The obscure hints in this message about a mushroom that would have unusual psychic effects were confirmed by the observations of Mr. Wasson in Mexico.[3] Now, while we had been attempting to join across the distance from Maine to Mexico by means of extrasensory perception, the first tangible, and unexpected, result of this reaching was in our hands. Was this the sacred mushroom we were seeking?

[2] No *Amanita muscaria* was found during the years 1956 and 1957 either.

[3] The mushrooms found by Wasson in Mexico are not *Amanita muscaria*. I merely point out that the mushrooms of Harry's writing and those of Mexico are both supposed to induce psychic effects.

chapter VII

WHEN I heard from Mr. Wasson on his return from Mexico in the early part of August 1955, I learned to my great disappointment that our experiment had not gone as planned. On the night of June 30, 1955, Mr. Wasson had attended the service of a *curandera* who performed the sacred-mushroom rite. It was his intention to direct the attention of the *curandera* to the laboratory in Maine in order to find out if she could get any intelligence about what was going on.

But the events of that night swept him into a course which he could not have anticipated and whose powerful influence did not allow him to carry out the plan. On this night Mr. Wasson received his initiation into the mushroom rite. The *curandera* offered him the grace of the sacred mushroom, and he entered into this unique experience. He has eloquently described it in the *Life* magazine article, and in his great work on the mushroom. He was carried into a visionary world where he saw things unknown painted in all the rich and vibrant color which can only arise within the mind, and not through the eyes, and which has no counterpart in the physical world that human beings ordinarily know.

The ecstasy of the moment and the strange night ritual in the Mexican village precluded all thoughts of directing the *curandera's* attention to the problem of getting intelligence from Maine. During the period that Mr. Wasson was in Mexico he did not have the opportunity to direct any of the mushroom ceremonies which he witnessed toward this objective.

Our sensitives in Maine received no verifiable intelligence about the situation in Mexico. This was pretty much what I expected, since

86   THE SACRED MUSHROOM

neither of our subjects had talent of the order required for such a prodigious feat. The unexpected by-product, the finding of a golden mushroom, was perhaps due to the fact that for ten successive nights our group in Maine had been tightly knit by the requirements of carrying out the experiment and by the common focus on the far-off situation in Mexico. The finding of the golden mushroom made the entire effort worth while, even though its original objective had not been fulfilled. We now had a kind of objective conclusion to the information which Harry Stone had delivered in a trance state on June 16, 1954, in New York City. On July 6, 1955, we had in our hands a golden mushroom which closely fitted the description given on June 16, 1954, and which was identified as a specimen of *Amanita muscaria*. After the discovery of this mushroom I at last had some reason to believe that the Ra Ho Tep performances of Harry Stone represented a rare and unique case.

I now turned my attention to the problem of finding more golden mushrooms so that we could begin to evaluate its effects on the human mind. It would require quite a few specimens in order to carry out satisfactory observations on many subjects.

Maine in the summer of 1955 was beautiful. It was what is known as a hot summer in Maine. This means that the temperature rose into the nineties. But it was not dry, because every few days or so there would be an accumulation of heat and moisture, and the clouds would gather, and rain and thunder would follow.

In the summertime I am usually quite busy because people choose that season to come and visit the laboratory. I spend a great deal of time entertaining such guests. Besides this, the press of work is greater because it is easier to get laboratory subjects in the summer than it is in the winter; therefore, I rarely have a chance to enjoy Maine in this season. My occasional escape from the laboratory and the work is to go out on the sea on fishing or exploration trips among the islands.

This summer, however, I spent a great deal of time on land, climbing the hills and beating my way through the forest looking for mushrooms. The discovery of the *Amanita muscaria* by Alice stimulated all of us to find more specimens. For the two weeks after July 6, 1955, when Alice found the first *Amanita muscaria*, the entire staff tramped the country for miles around the laboratory looking for more mushrooms, but the search was all in vain. It seemed that the only mushroom for miles around was the one that Alice had already found.

THE SACRED MUSHROOM   87

July 20 was my oldest daughter's birthday. In addition to her usual party, I had promised to take her on a hike up Bald Mountain behind our home. Jinny and our three daughters, Lanie, Ditza, and Illyria, and I started out with a picnic basket full of cold drinks and sandwiches up the mountainside. Illyria, the youngest, as usual either straggled way behind, or had to be lifted up and carried. There was a rather well-defined trail that went up the mountain used only by the Puharich family and assorted porcupines, woodchucks, and deer that traveled up and down. However, there were places in the trail that had become choked with fallen trees, or where the sharp needles of the juniper had grown into the path making it very difficult for the little ones to get by. Because of these obstructions we wandered off the trail and descended into a ravine. The bottom of the ravine was cool and moist and covered with moss everywhere. It had escaped the heat of the summer. We all sat down to rest and to have a drink of cold milk.

My children began to ask me questions about what I was doing at the laboratory. So I made up a rather fanciful tale about fairies and goblins and mushrooms, which aroused their interest. My story was designed to amuse the children, but also to get them to look for mushrooms. The children quickly got the spirit of the mushroom quest and insisted that we spend our time in looking for them. I described to the children how they should look for a bright shiny golden mushroom, which if it were present they couldn't miss, because of the way it stood out against the dark background of the earth or the greenery of the forest. With great excitement they leaped up and were off to find mushrooms.

Lanie led the way as we climbed up out of the ravine through the heavy overgrowth and over the moss-slippery rocks. About halfway up we were faced with a cliff too high and steep for any of us to climb, so we bore off to the right in order to get around it. There we ran into a small open clearing where there were some old birch trees, and some magnificent old oaks. Lanie, who led her mother along, was the first to get into this oak and birch enclosure. She squealed with great joy because between the biggest of the oaks and the birches Jinny and she had found a golden mushroom. We all rushed up in great excitement and stood around this treasure. The children jumped up and down with glee and wanted to know if this was the fairy mushroom that we were looking for. I saw that the mush-

88   THE SACRED MUSHROOM

room did indeed have the proper color and warts distributed over the cap and made a guess that it might be the *Amanita muscaria.*

I pointed out to the children that the mushroom was growing in the kind of spot where legend said it would grow. I pointed out the majesty of the oak. I told the children how the ancient Greeks and the ancient Druids had worshiped the oaks in the great forests of Europe. How they had believed that there was to be found either close to, or on, these monarchs of the forest a beautiful golden bough which one could use as a passport to a fairyland. I pointed out to them the queen of this monarch oak of the forest as being the magnificent white birch that soared over our heads. I told them how in far-off Russia the Russians called birches the maidens of the forest; and how the golden, or red mushroom in Russia (which is the *Amanita muscaria*) was often found growing in the birch forests. Then I told them how lucky we were to find the king of the forest, the oak; and the queen of the forest, the birch, with their little child growing at their feet, the golden *Amanita muscaria.* This story held them spellbound for a short while, but soon they wanted to take their treasure home. How were we to pick this beautiful child and bring it safely down the mountain to the house?

Lanie suggested that she would run back and get a small bushel basket and we would lift the mushroom and the earth beneath it and place it in the basket, and we would take it home where we could watch it grow. This seemed like a splendid idea, and Lanie went down the trail like a young deer skipping over the fallen trees and slashing through the sharp undergrowth. It seemed that she was back in minutes, so fast had she gone down and up. She was out of breath, but she had remembered to bring along a small shovel. Under the great oak, and under the benign influence of the birch, the five of us gently and carefully lifted our mushroom and placed it, earth and all, in the basket. Lanie felt that her birthday was a memorable one, because she had at last seen the golden mushroom of the fairies, and in fact she had shared in its discovery with her mother.

But searching for mushrooms was not always as pleasurable as the occasion I have just described. Usually I went out with two or three of my colleagues and we would systematically lay out an area according to a grid pattern and cover every square foot of it. It wasn't until July 24 that we had any success in adding to our supply of mushrooms. On this day Betty and Harry found three beautiful specimens of *Amanita muscaria* growing on the same spot where Alice had first discovered her mushroom.

This spot is very interesting because it describes many of the characteristic features of the natural habitat where this mushroom is to be found. The mushroom was found on a road which I have described running along the seacoast. This road was used at times by neighbors as a horse trail, and, consequently, it was sprinkled here and there with droppings. The association of the mushrooms and animal manure is too well known to repeat here. The spot where Alice had found her mushroom was about six feet off of this trail on the seaward side. It was a depressed shallow spot that had poor drainage and when the rain fell it tended to collect here. Therefore, with ample moisture the forest growth was luxuriant in this area. The mushroom-growing spot was in the center of a triangle formed by two oak trees and one birch tree.

Over the years the rotting leaves, logs, and sticks had piled up a rich spongy humus in this area. There was the salt of the sea air, the nutrition of the oak and the birch, the shallow basin where the deep soil was kept moist, and the ample shade which was broken only by the cleft of the road through the trees. This is what one would call a natural environment for the growth of the *Amanita muscaria*.

On July 27 it was raining lightly outside and the air was warm and humid. Periodically thunder would roll from the mountains to the west, and round their peaks the sky would occasionally be shattered by lightning. Remembering the folk tales and the ancient legends about the association between lightning and the growth of the mushrooms, it occurred to me that this would be a good time to go and search for them.

Betty, Harry, and I put on raincoats and rubber boots and sloshed off into the woods. We first went to the spot which was now affectionately called Alice's Cave to see if any more mushrooms had come. We were delightfully surprised when we found a new bud of an *Amanita muscaria* just pushing out of the moist soil. This was a wonderful opportunity to watch the growth of *Amanita muscaria*. Harry went back to the laboratory to get a camera and some color film while I just sat and watched this marvel of nature heaving up the soil and thrusting its proud head into the air.

I noticed that the mushroom looked like a small golf ball as it came through the earth. The entire nob of the mushroom was covered by a golden membrane which extended down below the ground

90 THE SACRED MUSHROOM

to the root level. I knew that as the mushroom grew further this membrane would burst, and it would split up into little pieces which in the mature plant would be the warts that we see on the cap. The lower part of the membrane would become the annulus or little necktie of the stalk of the mushroom. We were fortunate in being able to get a good series of colored photographs of this mushroom during the next twenty-four hours and were thus able to record every stage of its growth.

I spent every day of the following month, August, in scouring the woods but had no luck in finding any more mushrooms. By this time I had come to recognize those spots where it was likely that mushrooms in general, and particularly *Amanita muscaria*, would grow. I realized that a complex of circumstances was necessary before the mushroom could grow. In walking, I always looked first for either oaks, birches, pines or hemlocks. These were the trees most intimately associated with the *Amanita muscaria*. Secondly, I looked for the shady side of a hill or ravine where the moisture was everpresent, and certain types of rocky outcroppings where humus had collected that appeared to be favorable for the growth of the mushroom. But each time I found one of these promising constellations of growth factors I would be disappointed. The presence of these natural circumstances did not necessarily produce the *Amanita muscaria*.

In being so intense while on the lookout for this particular species of mushroom, I, too, began to feel, like the ancients, that the growth of this mushroom could not be accounted for entirely on the basis of natural circumstances. It is a fact, that even though *Amanita muscaria* is known in many lands, as far as I know, it is one of the few mushrooms that has resisted artificial cultivation in the laboratory. I myself had converted an old wine cellar in the basement of the laboratory into a small fungus farm. I used the spores from the specimens we had found and set up many different natural conditions of moisture, soil, and temperature in order to cultivate the mushroom. But I never was able to cause one single specimen of *Amanita muscaria* to sprout under these laboratory conditions. There is a legend among certain Mexicans that mule manure is particularly favorable for the growth of one of their species of sacred mushroom; but even this material did not help to produce growth in the spores in my laboratory.

In the latter part of August, I had to go to New York on business. I chose to go by car because I had several stops to make on the way. I was driving back from New York City on August 26, 1955, and decided to take a different route up the Hudson Valley rather than the U.S. 1 coastal route. I wanted to see some of the Berkshire country at the height of summer. I was crossing from the Hudson River Valley to Pittsfield, Massachusetts, on route U.S. 20, when I was stopped by a policeman just outside of Pittsfield. He informed me that the heavy August floods that had ravaged New England had washed out many bridges for the next fifty miles, and I would have to detour. He suggested that I continue north on Route 7, to North Adams, Massachusetts, and there to take Route 2, the Mohawk Trail, eastward. Since I had never been to this part of the country I was not too disappointed at this detour.

In leaving North Adams, which is in the Berkshire Mountains, one follows a very steep road up the mountain, and from its peak there begins a long eastward descent down what is known as the Mohawk Trail. Apparently the Mohawk Indians had used this notch in the mountains as a trail. The road is winding and narrow and has some rather exciting views overlooking the steep gorge below the road. There are very few places where one can stop a car in order to admire this scenery. I had come down the Mohawk Trail almost to the bottom and I had been keeping one eye on the magnificent views around me. I noticed that the birch, the oak, and the hemlock were clustered very heavily at certain points, and I noticed that small brooks and ledge seepage kept these areas fresh and moist. I decided that if I could find a place to stop the car I would get out and investigate one of these likely-looking places. As I was approaching Charlemont I found a small space to the right of the road where one could with safety park a car.

I got out of the car and found an old lumber trail striking off into the woods. As I walked along this trail I found many specimens of different kinds of mushrooms. It seemed that the time of the year was just right for mushroom growth. After a half mile of this trail I turned back, because I had not found a single *Amanita muscaria*. When I came to within fifty yards of my car I suddenly spied a huge *Amanita muscaria* shining with its golden luminosity through the shade of the undergrowth. I rushed for it in order to make sure that it was not an illusion. It was real. It was at least fourteen inches tall and had a cap fully eight inches across. The cap was covered with

92 THE SACRED MUSHROOM

several hundred small warts. This was the first luck I had had in finding a new source of the mushroom since early July.

I examined the area around the car square yard by square yard. All in all I found nine such large specimens of *Amanita muscaria* within fifty yards of the car. I extended my search for at least a half mile, in all directions from my car, but the only place where I found any mushrooms was the spot where I had by chance first stopped. This was too exciting to leave, so I decided to stay in this area for the next few days and continue my search.

I registered at a local motel, got some boxes of ice, and iced all my specimens so that they would stay fresh. I went into the village and talked to a number of the local inhabitants and asked them if they knew anything about the kind of mushrooms I was seeking. I did not find a single native who had ever seen the *Amanita muscaria*, or in fact had even heard of it in that area. So I got no guidance from these sources. The next day I was up bright and early and plunged into the woods to continue my search. I walked up and down this area for the next few days, and I believe that I must have covered five square miles minutely, but I had no further finds. My lucky bag of the first day was all I could come home with.

On September 1, I had gone to bed about one in the morning, which is early for me, exhausted from the long trip to New York. The night was foggy and there was a fine drizzling rain. This was just the kind of night in which I can fall into a deep sleep and nothing can wake me up. But I did not sleep well, evanescent dreams kept drifting through my consciousness. I arose finally and looked at my watch. It was 4:00 A.M., I had been in bed for three hours. My wakefulness was unusual; and my friends can testify to the fact that I am the laziest man in the world to wake up in the morning. I tried to recollect the dreams, but they were all fragmentary and I couldn't reconstruct any of them. So I tried to sleep again but I couldn't.

I got out of bed at four-thirty with the feeling that I had to do something, so I bathed, dressed, and ate, and at 6:00 A.M. drove to the laboratory which was about ten miles away. It finally occurred to me that what I should do was go out and look for mushrooms since the day was just perfect with the light rain and the fog. So I headed for the seacoast on the foundation property.

I went to Alice's Cave and found three new *Amanita muscaria* that had just budded through the ground. I began to feel keen, so I plunged on into the woods, following my instincts. I walked down on the low-tide rocks of the seashore and went north for about a quarter of a mile. To my left rose a rather sheer stone cliff about fifty feet high, and to the right was the sea. I slipped and slithered over the rounded rocks on the beach. But of course this was no place to look for mushrooms. So I decided to return to the shore above. I climbed up the face of the cliff and as I broke through the underbrush at the top I found myself in a clump of birch and oak, and there were two more specimens of the golden mushroom. These, too, had just budded through the ground.

Then I wandered north through the wet underbrush and found three more specimens just budding through the ground. This was the first time I had had an opportunity to watch so many *Amanita muscaria* coming up all at once. I didn't pick any of these specimens because they were too young, but just carefully noted their site in order to return later when they were full-grown. I found them all in the woods among oaks and birches.

Between the shore line of forest and the laboratory there was a large open blueberry field. I decided to explore this blueberry field where it adjoined the woods. And here again I found that there were a few specimens of the golden mushroom growing just at the edge of the shade of the trees. Now these mushrooms were growing in an open field. Their position was such that they were protected from the hot morning sun because the forest was to the east of them, and the sun would probably not hit them until about 11:00 A.M.

All in all, my early morning adventure was quite a success in that I had found seventeen specimens of *Amanita muscaria* in two hours. I got to the laboratory at 8:00 A.M. and announced my findings to my colleagues. I told them that all the specimens I had found were newly budded and probably would not mature for at least twenty-four hours. But we decided to check them sooner to see how they were coming along. At three o'clock in the afternoon, which was about eight hours after I had first found these mushrooms popping through the ground, I went back to look at my find. I was utterly amazed; the seventeen mushrooms had become full-grown, and, of these, ten were almost rotten from the heat of the sun and worm infestation. Here I discovered something which none of the books had mentioned about the *Amanita muscaria*. There is a certain small slug of a pale oystery color with two little horns on its head which seems to live only for an *Amanita muscaria* feast. These little slugs attack the mushroom from the base of the stalk and ascend the stalk

94 THE SACRED MUSHROOM

in its interior by eating their way along it. This, of course, cuts off vital nutrition from the mushroom, and so it collapses.

I had really learned something. These specimens, from the time I had seen them as little buttons just peeking through the ground, were completely full-grown and ripe within eight hours. I could only pick seven of the mushrooms that I had earlier marked, because the rest were crumbling as a result of being overripe and as a result of being eaten away by the little slugs.

I alerted the rest of the staff and we again covered the entire peninsula. After a three-hour search we came back empty-handed. There were no other specimens of *Amanita muscaria* to be found. The only ones present were the ones that I had uncovered earlier in the day.

In reflecting on this lucky haul of so many mushrooms on our property all at once, I concluded that the natural conditions, that is, temperature, season of the year, moisture, and other factors, were all optimal for the growth of the *Amanita muscaria*. I decided to return to my site on the Mohawk Trail in western Massachusetts to see if the mushrooms were growing there as well as they were growing in Maine. This was a three-hundred-mile drive, and I covered it rather quickly the next day. I stayed at this site for three days, and my luck was far better than anything I could have imagined. I found two hundred and thirty-five first-class specimens of *Amanita muscaria*. This gave me invaluable experience in learning where to find them, the different habitat in which they grew, and the various colors, shapes, and characteristics that they assumed under different conditions. This insight was to be a great help to me later on when I began to study the use of the *Amanita muscaria* by peoples in ancient times and in far-off places; in their metaphors and allusions to mushrooms, I could recognize readily references to the natural habitat of the *Amanita muscaria*.

chapter VIII

I now had an ample supply of *Amanita muscaria* with which to begin serious investigation. A number of problems had to be explored. The first one was the chemical analysis of the mushroom, the question here being the composition of the drugs present within the plant. The second problem was to apply the chemicals from the *Amanita muscaria* to human beings in order to understand its psychic effects. The third problem was to learn to handle the drug in such a way as to minimize or avoid its poisonous action.

The chemical studies with the mushroom confirmed what had already been found in the literature and did not turn up any new evidence. The mushroom like all plants is a composite of many chemicals. However, there are three chemicals in the *Amanita muscaria* that are of interest in their relationship to psychic effects: (1) muscarine, (2) atropine, (3) bufotenin.

Muscarine when applied to biological systems shows itself as a chemical whose effects can be divided into a number of phases. The initial effect of muscarine is to stimulate parasympathetic nerve endings, and this is observed in the vomiting and diarrhea usually following *Amanita muscaria* ingestion. That part of the sympathetic nervous system which is at the head and tail end of the human body is called the parasympathetic nervous system, and in general it is the one that is stimulated by muscarine.

It has been noted by observers in Siberia that the shaman who uses the *Amanita muscaria* is capable of great feats of muscular exertion and endurance. It is believed that a part of this prodigious ability for muscular exertion is achieved by the use of the mushroom and

95

96 THE SACRED MUSHROOM

that the particular chemical responsible is muscarine. However, after its initial stimulating effect, muscarine then acts as a poison and paralyzes the very nerves which it has stimulated. In this paralysis lies the cause of death from the accidental use of this mushroom.

The atropine present in the *Amanita muscaria* is commonly known as belladonna and was known to the ancients as the deadly nightshade. Atropine alone first stimulates the central nervous system and then paralyzes it. It causes hallucinations and may lead to convulsions. Curiously enough, atropine is a specific antidote to muscarine; that is, it counteracts the effects of muscarine on nerve-muscle endings which result in the symptoms described above. Therefore, large doses of muscarine can be counteracted by a proper dose of atropine.

Here we have one of those magnificent examples of the wonders of nature. In one plant we have a drug, muscarine, which has a given dangerous effect on humans; and the same plant produces atropine, which counteracts that dangerous effect. These two drugs, each with its characteristic powerful action, individually act in the human body to counteract one another. In this fact may lie the reason for the disagreement in the literature as to the poisonous effects of the *Amanita muscaria*. Some writers say that *Amanita muscaria* can be eaten with impunity. Other writers say that it is poisonous, and warn the reader to stay away from it. Some writers doubt whether it will kill flies, where others say it will. Could not all this conflicting evidence indicate that in any particular plant it is the balance between the presence of the atropine and the muscarine which determines the effect on the individual?

It is known that *Amanita muscaria* can be eaten with impunity if a number of things are done. In the first place, if the warts and the skin of the caps are removed the mushroom is safe. This would lead one to believe that the toxic principle is present in the warts and in the skin. Secondly, others have stated that the *Amanita muscaria* can be eaten if it is marinated either in salt or in vinegar, it being believed that this will counteract the toxic principle. It is also reported in the ancient literature that *Amanita muscaria* when used with milk has its toxic effects minimized. Thus we see that salt, vinegar, or milk do something, according to these traditions, to neutralize the toxic effects.

The third drug present in traces in the *Amanita muscaria* is bufotenin. This drug is also a secretion from the sweat glands of the

African toad (*Bufo-bufotenesis*). Now it is curious that an animal like the frog on the one hand and a plant like the mushroom on the other hand should produce one and the same drug. It has been pointed out by Wasson that there is a traditional association of toads and mushrooms, particularly in regard to the *Amanita muscaria*. This is proven by metaphors and legends built up around the mushroom as well as its many names in different languages, all of which associate the mushroom with toads. Here we have a rather unique wedding of ancient tradition with modern knowledge in that the same drug is indeed present in a certain toad and the mushroom; namely, bufotenin. In medieval times it was believed that one could be poisoned by toads, and there are many references in the literature about such poisoning, but until recent times nobody had any verification or certain knowledge that this toad, the European toad (*Bufo vulgaris*), was poisonous.

Bufotenin has an excitatory action similar to adrenalin. Bufotenin is also known as a drug that has hallucinogenic effects on the human. Dr. Howard Fabing[1] has reported on his experiments in which he injected human volunteers with large doses of bufotenin prepared from the frog. In these individuals there were marked hallucinations.

Thus we have present in the *Amanita muscaria* a unique combination of drugs, each of which alone is capable of producing marked hallucinations. However, they are present in this plant like a well-compounded prescription, and since one acts against the other it is difficult to predict the outcome of the effect on the human mind. This, indeed, complicated our study.

Our chemical studies of the *Amanita muscaria* were such that we prepared simple extracts from different parts of the plants; namely, the stalk; the pileus, or cap; and the pileus with and without its covering membrane. Separate extracts were made of the skin and the warts alone. These studies showed that the preponderance of the drugs which I have described are present in the membrane and in the warts. Therefore, in our studies we used preparations made entirely from the skin and the warts.

The studies were aimed first at evaluating the poisonous effects of the mushroom. To this end, I and other volunteers took the mushroom by chewing it, according to the method used in Siberia, in order to find out what the effects were. It must be said at the outset

[1] Howard D. Fabing, *Science*, May 18, 1956, Vol. 123, pp. 886–87.

98   THE SACRED MUSHROOM

that the effects for each individual represented a different constellation of symptoms. But, in general, these are the symptoms noticed by most of the subjects.

The first reaction noticed was that the skin would feel hot in some places and cold in other places at the same time. When a subject was observed, it was found that his skin was blotchy, red, and discolored in some areas, and blanched and white in others. For example, one ear might be flaming red, while the other one would be an ivory white at the same time.

Some individuals noticed a disturbance of vision primarily in the form of blurring. This probably was an effect due to the atropine. Atropine tends to dilate the pupils if not counteracted by an adequate amount of muscarine.

The third effect was rather an objective one in that almost every individual experienced a lowering of the pulse rate, usually from the normal level of seventy or eighty beats per minute to fifty or sixty beats per minute. There was a slight lowering of the temperature in most individuals.

Some individuals reported a hypersensitivity to touch, to light, and to sound. Other individuals were not affected by such hypersensitivity at all. Some individuals noticed increased strength and endurance. For example, one subject had made a study of how long he could hold his breath. His previous record time had been one minute and thirty-two seconds. Upon chewing a small piece of *Amanita muscaria* he found that he was able, with ease, to hold his breath for two minutes and thirty-two seconds as a maximum.

Interestingly enough, none of the normal subjects, and thirty-five were studied, experienced any noteworthy psychic effects from the mushroom, either in the form of hallucinations or mental disturbances. There was a reason for this, in that I kept the dosage purposely minimal in all cases. I wanted to study the effects of minute doses of the mushroom. The literature had already supplied us with ample knowledge of the effects of massive doses of *Amanita muscaria* in the way of poisoning, drunkenness, disorientation, and hallucination.

Perhaps the most curious effect of the mushroom, and by this I mean the strongest effect, was the symptoms noticed by the subject on the day following the use of the mushroom. In all cases where a sufficient dose had been given these individuals reported an unusual sensitivity of the gustatory and smell senses. The first thing they no-

THE SACRED MUSHROOM   99

ticed was that there was a perpetual bad taste in their mouths. There was nothing that could be done to get rid of this. Secondly, everything seemed to smell unusually foul. It was as though every smell was offensive. This delayed effect was certainly a very interesting one. None of our subjects reported headaches or any other unpleasant symptoms on the following day, because the dose had been kept within small limits. There was also an urgency to urinate without much issue.

As I have stated, of all the normal subjects studied none reported any remarkable effects in the way of excitation or depression of the psyche. There were two exceptions to this generalization. Out of the thirty-seven people studied, two were individuals who were proven by laboratory tests to be sensitives, that is, they could actually demonstrate extrasensory perception. These were Harry Stone and Peter Hurkos.

At the time of the first experiment with Harry we had in our possession only a few specimens of *Amanita muscaria,* and because of its poisonous nature, I was not inclined to be in any hurry about human experimentation. I knew far too little about how to handle it. On August 7, 1955, Harry was giving a demonstration of telepathy for Aldous Huxley. In the middle of the demonstration Harry spontaneously slipped into a deep trance. At the moment I felt that he had ruined my neat little laboratory demonstration. But in the presence of Aldous and myself, he entered into a dramatic sign-language demonstration whose meaning this time was quite clear. The Ra Ho Tep personality insisted on having the golden mushroom brought to him. I could not escape the urgency of this appeal. I had to go and get one of my precious golden mushrooms. I brought it back to the laboratory and placed it in front of the deeply entranced Harry. The Ra Ho Tep personality became ecstatic over it. Then for the first time I saw the secret details of how the mushroom was to be used. Harry applied the mushroom himself on the tongue, and on the top of his head, in ritualistic fashion. Five minutes after he had completed this remarkable demonstration he woke up.

He looked at Aldous and myself and weakly asked if I had given him some alcohol. I assured Harry that I had not given him any alcohol, because it was my intention to observe how he would behave if he did not know he had taken the golden mushroom. A few minutes later he began to stagger around as though he were heavily intoxicated with alcohol. The symptoms became alarming. The only thing to do was to counteract the effect of the mushroom by giving him atropine. While I busied myself with drawing some atropine into a syringe, Aldous watched him closely.

Harry was smoking a cigarette at this time. Aldous called me urgently and pointed out that the cigarette was burning the skin between Harry's fingers. Harry was completely unaware of it. I told Aldous that this anesthetic effect was to be expected from an overdose of *Amanita muscaria,* and that I would soon give him the atropine.

But before giving Harry the atropine I asked him how he felt. He mumbled that he felt very, very drunk. Then he looked straight ahead and said that he felt he could see through the wall of the laboratory. He said that everything seemed so clear on the other side of the wall. I asked him what he saw, and he gave me an accurate description. But I also knew that he had prior knowledge of the other side of the wall and this could well be imagination. Therefore I delayed giving him the atropine in order to do one quick test of his seeming clairvoyance. I hastily blindfolded him, urging his co-operation, and placed him before the covered MAT test. I begged him to try to do one test. Aldous and I watched him closely. His hands fumbled over the picture blocks for about a minute. He just couldn't seem to make his hands follow his will power.

I spoke sharply and commanded him to begin the test. He pulled himself together and completed the entire series of matching ten sets of pictures in about three seconds. He literally threw the two sets of picture blocks together. I took the cover away from the blocks, and was amazed to find that he had scored ten correct matches. The statistical odds against getting this score by chance alone were such that he would have had to do this test a million times before such a result would occur once. This was the most remarkable demonstration with the MAT test that he had done up to this date.

Now there was no more time to be wasted. I had to stop the drug action. I gave Harry a large dose of atropine. I removed by lavage the remaining particles of the mushroom. The treatment was effective and within fifteen minutes he was normal again. But I must say that he was thoroughly shaken by this experience. It was only then that I revealed to him the entire course of events. He was too stunned to believe me. He questioned me many times in the following days trying to grasp what had happened, and to understand the powerful effect of the mushroom.

The other subject who was unusually sensitive to the effects of *Amanita muscaria* was Peter Hurkos. The first occasion on which he was given the mushroom was October 19, 1956. I administered the mushroom to Peter, and to myself as a control, at 10:00 P.M. We both sat alone in the laboratory in silence, making notes of our own reactions. I, of course, was also making notes of Peter's reactions.

Now Peter is the type of human being who is constitutionally incapable of sitting in one place for more than five minutes. He is that restless. After I had administered this drug to him, I was surprised to find that Peter sat quite peacefully for an hour. I then spoke to him and found out that for the first time since I had begun to study him, which now was a period of six months, he had slipped into a light trance. He was not asleep in any normal way in which I can understand sleep. He was definitely in trance. His eyes were wide open. He seemed to be looking off into the distance but oblivious of everything in his immediate vicinity. He sat in this position for three hours. For a restless man like Peter Hurkos this was indeed phenomenal. At one point during this three-hour trance he wrote a statement on his own observation pad in the Dutch language. In translating these lines later, I found that they were statements which could be called precognition. One of them related to a personal event which was to happen seven months later, according to the date that he wrote down. Without describing this event I can say that this prediction made seven months in advance was quite accurate.

When Peter awakened I waited for him to speak to me first. I, myself, had had no noteworthy effects from the *Amanita muscaria*. I had been quietly observing Peter Hurkos for the past three hours. Peter first looked at me and wonderingly asked what time it was. I told him that it was one o'clock in the morning.

"Do you mean to tell me that we have been sitting here since ten?"

"Yes," I answered, "you have been sitting in that spot for three hours without moving."

"How is that possible?" he said. "I have never done that before in my life."

"Peter," I said, "I don't know how it's possible. All I can tell you is that it happened. Tell me, Peter, what did you experience during that time?"

102 THE SACRED MUSHROOM

Now, for the first time, Peter seemed to be fully aware of his surroundings and what he was doing.

"Andrija," he said, "I have seen things which I don't believe I could ever describe to you in a million years. I was not here in this room. I don't know where I was. But I was in some far-off place of indescribable beauty. The colors, the forms are beyond description. The only way I can give you an idea of what I saw is that everything around me here is filthy, dirty, and horrible. It looks so ugly here that I hope you don't give me that mushroom too many times; I might not want to come back."

"But, Peter," I said, "this is hard to believe. As far as I could tell, you were just sitting there asleep even though your eyes were wide open."

"Well, I'll try to describe it as best as I can. In the first place, I don't know where I was, but I was somewhere outside of myself. A woman came to me. I don't know who she was because she would never turn her face to me, and it could all be in my imagination. This woman guided me and took me I don't know where. I don't know whether I was walking or flying or what. I never had this before so I can't describe it. We came to a land. I didn't see trees. I can't say that, but I did see flowers. I can't describe them, they were so beautiful. I saw houses; there were many, many houses. The only thing that I can tell you about the houses is that they all looked like cupolas, they were like beehives. They were round and all with beautiful colors. I know that where my mind was is real. Its beauty is beyond description. And this world is so ugly. I'm sorry I came back."

Now this entire episode surprised me greatly. In the first place, Peter is not the kind of person whose imagination would drift in the direction of rhapsodizing beauty. His imagination is such that it would color something he had objectively done in rather glowing terms; or perhaps describe his ordinary human exploits with a bit of braggadocio. I have never heard Peter describe anything for its beauty's sake and enthuse over it.

Peter had come to America six months before, and his power of expression in English was limited. He was still learning the English language. My main impression was that he had mentally experienced a wonderful hallucination. He did not report anything that led me to believe he had been out of his body in the sense in which I have described it earlier. His description of being in a strange land was very much like his usual experience of having an impression by extrasensory perception of a distant scene on earth. If, for example, he sees a distant scene of a city or a house by extrasensory perception, he describes it as though he were there, but in actuality he is describing what he sees in his visionary mind, and there is no true experience of a separation of body and consciousness.

I have had the opportunity to repeat this experiment a few more times with Peter Hurkos. I could not do it too many times because his reaction was always the same. He did not want to plunge too deeply into this world of beauty he had discovered because, as he said, it was too unpleasant coming back. I shall report one more experiment done on August 23, 1957. Hurkos had been administered the preparation of the mushroom. He slipped into a semisleep state in about twenty minutes, and this time he began to talk. He said two things which bear repeating. In the first instance, he saw quite clearly what he called "a miracle in the sky." When asked what the miracle in the sky was, he was not capable of giving it finite description. But these are the words that he used: "There is going to be a miracle in the sky. It is coming. I cannot tell you precisely what it is, except that I see it as an earth-ball. It is in the sky, and everybody in the whole world can see it." When asked if this meant a planet, he said, "No." When asked if this meant a comet, he said, "No." I asked him of all the possibilities I could think of in the way of natural aerial phenomenon. I even asked him if this was going to be a flying saucer. He said, "No," to this. He stuck by his description that there would be in the sky "an earth-ball," and that everybody in the world could see it. I must say that to me at this time the description was puzzling.

The second thing he stated was that he saw an event which was to occur on September 27, 1957. This statement was very good for experimental purposes because he not only described an event that he saw coming, but he gave it a precise date. He would be right or wrong on the basis of the event's occurring on a certain date. What he said is as follows: "I see high government officials coming to this laboratory to talk to us. They will come on September 27 of this year. They won't believe me."

These are the two outstanding things that Peter Hurkos said under the influence of the *Amanita muscaria* in a semitrance state on August 23, 1957. Now, in reference to the first statement which is very vague, it is difficult to relate it to any specific event. The only event that seems to bear any resemblance to the words that Peter uttered was, of course, the launching of the Russian earth satellite on

104   THE SACRED MUSHROOM

October 4, 1957. But Peter himself feels that this is not the event in question. It is yet to come, he says.

The other event which Peter said would happen was easier to monitor. On September 12, 1957, a military friend of mind phoned from Washington with rather startling news. He said that he had been talking to some colleagues about our research in Maine and two officials had expressed an interest in visiting the laboratory. He told me that one of the men, a busy general, had picked a date to come to Maine. The date was September 27, 1957. My friend was calling now to find out if this date was acceptable to me. I said, of course, that date was all right with me; but how did he come to pick this particular date? He said that the general was going to finish an important mission in the West on September 25 and would be free on the twenty-seventh to come to Maine, and this had dictated his choice of the date.

This episode is only interesting in the light of what Peter had said, so I awaited developments. On September 16, my friend again called me from Washington and said that some hitch had developed, and that the projected visit of the general and the colonel had been called off. I asked him the reason for this change of plan. He informed me that there was some compelling security reason unknown to him which made it undesirable for military officials to express an interest in our kind of research, and especially to pay us a formal visit. The military group later cancelled their projected visit entirely.

Now this instance illuminates something that I have noticed a number of times in my researches. A sensitive will predict some event as occurring in the near future; the event does not occur, therefore, the sensitive is said to be wrong. In the case which I have just described, Hurkos picked a date and associated it with government officials and a visit to Maine. As far as I knew at the time when he made this statement there was no connection between myself and anybody in the government which would lead me to believe that such a trip was in prospect. However, unknown to Peter and myself, a friend had made arrangements in Washington for such a trip which was planned for September 27. Then other circumstances changed the course of this plan; the event did not come off. It is important to note here that Peter saw an event which certainly was in the making. He did not, however, foresee that the event would not go through as he had interpreted it. He only saw a part of what was happening.

It might be added that this sort of case also illustrates the fact that the workings of the human mind are such as to bias the clear cognition of things perceived, either by means of sensory perception or extrasensory perception.

It must be emphasized that this degree of extrasensory perception shown by Hurkos cannot be entirely credited to the effects of the mushroom. The mushroom was responsible only for placing him in a light state of trance. I have seen Hurkos demonstrate just as good or better examples of extrasensory perception without the use of the mushroom.

That normal human beings do not show any remarkable psychic effects or extrasensory perception from the mushroom is quite clear; but on the basis of my limited observation it seems that sensitives are more responsive to the psychochemical effect of *Amanita muscaria*. This effect may be due to a greater sensitivity on the part of sensitives to the chemicals present, or it may be a greater psychological susceptibility to the things that the mushroom is supposed to be able to do. It is my opinion that future research will show that *Amanita muscaria* has a selective psychic effect on genuine sensitives by increasing their powers of extrasensory perception. This presents a razor-edged problem that can only be cleared up by a great deal of research with many sensitives, and different psychochemical drugs, including controls by means of placebos.

In spite of this unresolved question, some problems had been cleared up by the mushroom studies. We could be quite sure that the *Amanita muscaria* did not have the powerful hallucinogenic effects of the Mexican group of mushrooms as reported by Wasson.[2] The effects noted on normal humans who had taken the *Amanita muscaria* were well-known physiological reactions and not in any way remarkable in regard to psychic effects. The effects noted on Harry Stone and Peter Hurkos are in a class by themselves, and one cannot dissociate the mushroom action from the already present remarkable talent for extrasensory perception.

[2] The difference may be a question of dosage. The drug present in one of the species of mushrooms, *Psyilocybe mexicana*, discovered by the Wassons has been isolated, and named psilocybin. *Time*, June 16, 1958.

chapter IX

I now turned to an evaluation of Harry Stone's role in the mushroom drama that had unfolded. It must be remembered that the finding of the golden mushroom was an event quite independent of what Harry had said or done. The fact that Alice had prepared the message while in a hypnotic state, and that this message preceded by two days the discovery of the mushroom, clearly placed this event out of any manipulation that Harry Stone could have achieved. In spite of the independence of these events, i.e., the finding of the mushroom and Harry Stone's role, the two events were intimately linked in the final result. This was perhaps the most satisfactory evidence that I had had to date that there was something outside of the personalities of Harry Stone and Alice Bouverie that was responsible for what had actually happened. In spite of this realization it was important—I am sure to the reader also—that Harry's personal role be thoroughly evaluated.

My relationship to Harry can be divided into four phases. From June 16, 1954, to June 1, 1955, was the period in which I knew about him mostly through Alice, because I, myself, met him on only a few occasions.

From June 1, 1955, until September 1955 was the period during which I observed Harry's deep trances while he was working daily in the laboratory. The last time he showed any signs of the Ra Ho Tep personality in my presence was on September 8, 1955. On February 7, 1956, he went into deep trance, but not in my presence.

From September 1955 until November 1956 he continued to work

108   THE SACRED MUSHROOM

at the laboratory as a subject in telepathic experiments. During this period (with the exception noted) there was no manifestation of his trances. He left the employ of the laboratory in November 1956 but continued to live near the laboratory, working at his sculpting. Our friendship continued during this period, and I saw him frequently.

During the period when Harry was under the observation of Alice in New York, she never noticed anything that would lead her to believe that Harry was using his trance performances for ulterior motives. After Harry came to Maine and I had him under my daily observation, I, too, could find no ulterior motive which would lead me to believe that his trances were either conscious or unconscious frauds. During this time I found him to be a loyal, hard-working, and honest chap. His sole motivation in life seemed to be toward bending every effort to become a good sculptor.

His prime means of expression seemed to be through his hands. He was a highly skilled craftsman and a good sculptor. I don't believe I have ever seen Harry Stone read a book. I have seen him thumb through art magazines, looking at pictures, but I do not recollect that he ever bothered to read the text. I have never seen him take any interest in reading or discussions concerning ancient Egyptian life. In fact, the only thing I ever saw him read was the newspaper, and in this reading he seemed to be primarily interested in international politics, especially the great struggle going on between Russia and the United States. Even his political interests were not very well defined, and I could only surmise that he had a sort of Dutch point of view in these matters, conditioned largely by his war experience. He certainly did not impress me as the type of personality that one usually finds involved in a confidence game.

Harry, in general, tended to ignore both his trances and his work as a telepathy subject in the laboratory. He was very embarrassed if somebody asked him about either of these two phases of his life. He preferred to be known only as an artist and as a sculptor. He certainly did not use his gifts in the field of telepathy, or the infrequent trances, as a means of attracting attention or becoming a center of interest.

From September 1955 to February 7, 1956, his trance tendency became more and more feeble. The last clear message came on September 8, 1955. After this he did go into light trance a number of times and attempted to communicate both verbally and in writing,

THE SACRED MUSHROOM 109

but neither the spoken words nor the written characters came through in any understandable or legible form. It was apparent that this phenomenon was leaving him and he could no longer act as an effective instrument under trance. His last trance communication was on February 7, 1956, during which time he covered a page with very poorly written hieroglyphs which are completely illegible.

With the decline of his trances, there also occurred a rather slow process of loss of motivation on his part. He became less active in his sculpting work, and in fact at times lost interest in it. His work in the laboratory became poorer and he seemed to have very little inclination to go ahead with it. He became personally irritable and at times erratic in his loyalty to friends. It was at this time that he became unpleasant at times to me, and even to his good friend, Alice. He seemed to resent what he thought was our patronage. This loss of motivation and interest and occasional wavering of the bonds of friendship with others gradually became worse. Certainly such behavior was not calculated to ingratiate himself with anybody at the laboratory or to gain him any favor.

A number of factors brought this instability to a head in April of 1956. While arrangements were being made to bring Peter Hurkos to the laboratory, Harry seemed to feel a great deal of insecurity about Hurkos' impending arrival in Maine. I suppose he didn't feel that he could compete with a famous and well-known sensitive like Hurkos because he had so little confidence in his own ability. He also had some fear that the close association with Hurkos might be revealed to his friends in Holland and it would become known that he was working as a sensitive in a laboratory. This he did not want to happen.

It is difficult to understand his feelings at this time because he kept them to himself. But one day while I was in New York on business I received a phone call from the administrator at the laboratory saying that Harry had suddenly decided to leave permanently. This distressed me very much because I did want to complete my study of Harry, and I did want to use him in telepathy experiments with Hurkos. In the case of the projected experiment with Hurkos I felt that it was a rare opportunity to have two Dutchmen, both of whom were sensitives, in the same laboratory. This would give me a new opportunity to study telepathic interaction between two men who I had every reason to believe might serve as an effective team.

I managed to intercept Harry at La Guardia Airport in New York City as he was changing planes to go to the West Coast. On April

12, 1956, I spent the entire night with him, trying to find out why he had suddenly decided to leave us, especially without any warning. The best I could get out of him was that he was tired of being a sensitive, and didn't relish the idea of working with a famous sensitive like Hurkos. The conclusion I came to was that he had just been overwhelmed by a compulsion to leave, and had acted without much premeditation. He said he was very sorry to have put me in an embarrassing position, and especially for interfering with the plans I had made for him and Hurkos to work together. He made no new demands on me when he agreed to return to Maine. It was certainly not his purpose to bargain with me for any better salary. In a few days he did return to Maine and settled down both in laboratory work and in sculpting.

With the arrival of Hurkos in Maine a decided change came over Harry. He took an interest once more in everything he was doing. However, a more important change came over him in that he became interested for the first time in the phenomenon of extrasensory perception. In Hurkos he saw a man who had full confidence in his ability and who could demonstrate extraordinary things at will. For the first time Harry came to believe that there was some reality to extrasensory perception. Up to this time not only did he not have confidence in his own abilities and avoid discussion of them, but he had also doubted my own interest in the phenomenon of extrasensory perception. When I would tell him that he had achieved a good score in a telepathy test I could see that he doubted it. He seemed to take the attitude that I was just saying this in order to encourage him. But all this changed with the advent of Hurkos, because now he saw that independent of himself, and independent of anything that I did as an observer, Hurkos could, indeed, do unusual things. In this sense, Hurkos gave Harry a confidence that he never had before.

Peter Hurkos must be described at this point. He is a huge man, six feet, three inches tall and weighing over two hundred pounds. He is full of energy, enthusiasm, and confidence. This was a sharp contrast to Harry's slender build, quiet manner, and introverted character. Curiously enough the two got along famously and became one of the best telepathic teams that I have ever had the opportunity to work with.

Hurkos' special ability is psychometry. I have had ample opportunity over a period of a year to study this gift in Hurkos. It took me about two months before I became thoroughly convinced of his

THE SACRED MUSHROOM   111

phenomenal accuracy in psychometry. For example, I gave him a handwritten letter sealed in an envelope sent to me by a woman who was over two thousand miles away from the laboratory at the time of the reading. This letter was handed to Hurkos without any comment. He promptly described the person, whom I had never seen up to that time, and whom he had never seen. Later on, I found that his description of the person was remarkably accurate as to physical and temperamental characteristics.

But the remarkable part of this reading was that he described in great anatomical detail a disease that this person had. He made a drawing to illustrate what he was trying to describe, since he did not have the command of medical language with which to verbalize what he saw. He drew an outline of the uterus and Fallopian tubes of this woman, and precisely outlined the anatomical abnormalities present. Furthermore, he said that the disease process was such that it would require an operation in order to clear it up. At the time of the reading I had no way to check the accuracy of his vision. However, seven months later the person whom he was describing became acutely ill, entered the Mayo Clinic for observation, and was operated on for the disease which Hurkos had described. The picture Hurkos had seen in his mind of the disease process in the Fallopian tubes was confirmed exactly by the pathological findings during surgery. This is only one example of hundreds of similar psychometric readings which I have seen Hurkos do, and which I have had fully checked afterwards. My confidence in Hurkos' ability was slowly and surely built up by such observations.

Because my confidence in Hurkos' accuracy in psychometry had been strengthened by my observations during May and June of 1956, it occurred to me that I could use him to check on Harry's Egyptian trances. My plan was to take one of the writings in Egyptian hieroglyphs that Harry had done under trance, seal this piece of paper in an envelope, and hand it to Hurkos for a psychometric reading. Because Hurkos was now thoroughly familiar with Harry's personality, it appeared that he might most readily give a description of Harry while doing the psychometry. If the trance writing of Harry really represented the guiding influence of an independent personality then Hurkos might get information, not about Harry, who had actually done the physical writing, but information about the alleged control personality. In either case, it seemed to me to be an experiment

112 THE SACRED MUSHROOM

well worth trying. I came to the laboratory in the morning on July 18, 1956 with the intention of carrying out this experiment.

In the laboratory I found Hurkos having a cup of coffee by himself and looking very disturbed. So I sat down and had a cup of coffee myself, and asked him what the trouble was. I had never seen him so agitated. He began by telling me that he had absolutely no faith or belief in the existence of spirits. The gift that he had, he felt, was solely the product of his own mind and sensitivity. He gave no credit whatsoever to any other influence in this work. He went on to say that he had never had any experience in his life which led him to believe that spirits were real. He continued at great length along these lines until finally I had to interrupt and say, "Why are you trying so hard to convince me that you do not believe in spirits?" Hurkos' reply was surprising.

"I know you won't believe this, Andrija, but last night my wife asked me to go downstairs and get her a sandwich and a cup of coffee. I came downstairs into the main hall and turned on all the lights along the way and went into the kitchen and brewed a pot of coffee and made a sandwich. I put them on a tray and started walking back through the main hall. As I entered the middle of the great hall I saw on my left, about twenty feet away, coming through the closed door, a large luminous mass. My first reaction was that there was a searchlight shining on that wall from the opposite end of the hall. So I looked to the right and saw nothing. I looked to the left and saw that this luminous ball was still there; and then it suddenly moved toward me, and I swear that it lit up the whole room, it was so bright. And as it went by me I could feel a cold breeze on my face, and then it left the room by the window on my right. Now I swear, Andrija, I don't believe in spirits, but I have never seen anything like this. I was so frightened that I spilled the coffee all over the tray and could hardly move from the spot. I then ran up the stairs and sat down in our room. Maria tells me that I was white as a sheet and that I couldn't say anything for ten minutes. Believe me, I was scared! I don't know what it means, but it really frightened me out of my wits."

"Peter," I said, "I don't doubt that you had this experience. I don't know what it has to do with spirits, because as far as I can tell, you didn't see any spirits; all you saw was a luminous ball, and it felt cold to you as it passed, is that right?"

"Yes," Peter said, "that's all I saw, but somehow it made me remember everything I had heard about spirits, and if I ever walked into a ghost I imagine that's how I would feel."

Our conversation ended here and I reminded Peter that we were scheduled to go into the laboratory and do some work. He finished his coffee, still talking about the startling experience of the previous night.

The experiments for the day were in psychometry, in which I had Peter work on a letter that had been sent to me from Formosa and a letter from a distant university. My procedure was to have Peter do a reading on these letters, after which I would write to the people who had sent them in order to get verification for his statements. The third test object was a photo which had been sent to me from Mexico. Peter psychometrized these three sealed objects in order, and I duly recorded his statements. For the fourth test I took him into a completely dark room and handed to him in a sealed envelope the Egyptian hieroglyphs reproduced below, which Harry had written in New York on December 9, 1954. (See Appendix 2, DRAWING No. 9, for translation.)

These are the statements that Peter Hurkos made of the impressions he got from handling this piece of paper: "All the people walk in women's dresses. No shoes are worn. A nice climate. I see a lot of sand. I see round doors. Strange, I hear a foreign name. It sounds like Kama. Hey! they write not in letters but in inscriptions, in pictures. The houses are very strange. From this place comes very strong powers, and this is from many years ago. It is a strange world. The people are very clever with their hands. There are no lawyers here, and they kill right away for an injustice. I see a dog on the wall. I see bows and arrows. I see a young girl dead, dead early in life. The body is well. I hear a name, Nakat-nile, she looks like a princess. There is much water with this girl. Her dresses are all covered with little figures of hands, and she wears earrings. I see palm trees in this land. Now I see a great warning hand in the sky and it is like the rainbow.

And I see it pointing into a pyramid, and the hand says: 'If you go in there you will never be happy again.' I don't know what is in this letter, but I'm sure that it must be thousands of years ago."

Then Peter said that he could draw the scene which he envisioned in connection with the hand pointing out of the sky. He made the following drawing:

To me this was a remarkable reading. After all, my purpose was to test the authenticity of Harry Stone's writing done under trance. The outcome of such a psychometric reading could only be:

1. That he got nothing, or would be completely wrong.
2. That he would describe Harry.
3. That he would get some picture connected with Egyptian life.
4. That he might get a mixture of all three.

However, his reading was clearly and exclusively that of an Egyptian scene. In fact the word he heard which I have spelled as "Kama" is indeed close to the ancient name for Egypt, Kem.

It was also interesting that he in no way associated this psychometric reading with Harry, whom he now knew very well. This was in its strange way one of the strongest bits of evidence that I had been able to get to date, that Harry had not fabricated this material from his own experience in any way, either consciously or unconsciously. In the light of such evidence, one must look to a source outside of the personality and experience of Harry Stone in order to understand its import.

The next morning, July 19, 1956, I again arrived at the laboratory to find Peter in the kitchen having a cup of coffee. He was still

THE SACRED MUSHROOM 115

talking about the luminous mass which he had seen two evenings ago. I quizzed him all over again as to what he thought it meant. He admitted that he had no idea what it meant, but that it had made an indelible imprint upon his mind. While I was talking to Peter the phone rang and I answered it. It was Alice Bouverie's son calling from New York City. He was very taut, and tersely announced: "Mummy is dead." I couldn't believe my ears, and asked him to repeat what he had just said. He repeated: "Mummy is dead, we found her in her bedroom this morning. She had apparently died sometime during the night. I thought you would want to know." I was so stunned that I couldn't find the right things to say. I finally asked him what had been the cause of death. He said he did not know, but the doctor who had examined her had the idea that it probably was a stroke. She had been in perfect health when she went to bed.

I put down the phone and walked back to sit with Peter. "Peter, that phone call was from Alice's son in New York. He said that Alice died last night." Peter looked at me numbly disbelieving, and said, "Now I think I know what my vision of two nights ago means."

Alice's sudden death remained inexplicable. The coroner described death as due to natural causes, but could not define any single pathology that could be responsible for such sudden death. I remembered that many years ago I had been faced with a similar case. While on night duty at the Permanente Hospital in California, a man, age forty-four, had come into the emergency room, saying that he was very ill and wanted to be admitted. I examined him for an hour but could not find any specific illness that would justify hospitalization. Since the hospital was filled to capacity, I could not in good conscience admit him as a patient. I explained this to him, but he still insisted that something was wrong and that he wanted to be admitted into the hospital. I called in one of my associates to make an independent check of his condition. My associate, after a thorough examination, agreed with me that there was no detectable ailment that would justify hospitalization. We both explained this to the patient, but he was still strong in his insistence on being hospitalized. Since there were no beds available I had to be firm in my point, and suggested that he go home and rest in bed and keep in touch with me.

Reluctantly he assented to this. I conversed with him for about fifteen minutes in order to find out more about the emotional factors which were bothering him. He was in the midst of telling me about

116 THE SACRED MUSHROOM

the kind of problems he had when he gave a long gasp and stopped talking. I could see immediately that he appeared moribund. I checked his pulse and blood pressure and found both absent. In the hospital I had all the facilities necessary to resuscitate him. I administered oxygen and artificial respiration, and after a few minutes injected adrenalin directly into his heart in order to get it beating. My colleagues and I worked over him for half an hour. Death was firm in its grip; there was nothing we could do to bring him back.

Since this was an inexplicable death, it became a coroner's case, and a very careful autopsy and examination was made the next day. The coroner could find no physical cause of death. He, too, made out a report that death was due to natural causes, but he had no idea as to what those causes were. It was my feeling that this, too, was probably the case with Alice. It was a tragic ending. She had been a great and good friend to all of us in Maine.

In the week following her death the question as to the association between Peter's vision, his psychometric reading of July 18, and Alice's death was before us. But we could not clearly resolve the problem posed. There were factors involved beyond our comprehension. We sadly turned back to research work.

In spite of the strong confirmation that Hurkos' psychometric reading had given me in regard to the authenticity of Harry's writing, there still remained questions which had to be answered. It occurred to me that perhaps Harry somewhere in his life had come into intimate contact with certain written Egyptian material, and this had been buried in his memory storehouse at a level far too deep to reach by ordinary interrogation. The only way to clear up this question was by the use of hypnosis. But how to do this hypnosis so that we would get a clear-cut answer created quite a problem. I knew from my work in extrasensory perception that there could be considerable telepathic leakage between the hypnotist and his subject. Therefore it was out of the question for me to try to hypnotize Harry; I knew too much about the Egyptian material and therefore might tend to influence him. If I got another hypnotist I could not tell him about the Egyptian material which I was trying to assess, because he, too, would tend to influence Harry. An opportunity finally did arise of having Harry hypnotized with the least possible conscious influence being exerted on him.

We had scheduled at the laboratory for the end of September 1956 a telepathy demonstration for an investigating committee

THE SACRED MUSHROOM 117

from M.I.T. We were going to demonstrate laboratory telepathy between Peter Hurkos and Harry Stone. Since this was quite a strain on both of them, it appeared desirable to get one more good telepathic subject, so that the entire load would not be thrown on Harry and Peter. There was a Mr. Jochems in California whom I had never met but who I had good reason to believe was a competent telepathist and an excellent hypnotist. I decided to bring him to the laboratory in Maine at the last moment on the pretext of using him as a subject in the telepathy experiment. And knowing full well how parlor hypnosis can be touched off spontaneously, I felt confident that when Mr. Jochems got to the laboratory, he would sooner or later engage in some hypnosis. It was my hope that, by leaving him alone with Peter and Harry, such a situation would arise and he would attempt to hypnotize them. If this occurred, then my initial objective would be achieved. This in fact did occur shortly after Mr. Jochems' arrival.

It seems that on the second night Mr. Jochems was at the laboratory, Peter and Harry had some friends over for a little party and induced him to do some hypnosis. He not only hypnotized several of the people present, but was able to hypnotize both Harry and Peter successfully. Now, as far as I knew, neither Harry nor Peter had been hypnotized before. When I heard of this successful hypnosis the next day, on my arrival at the laboratory, I was quite pleased at the result. The only suggestion that I made to Mr. Jochems was that he should try to hypnotize Harry again and implant a code word with him, so that when he wanted to hypnotize him again, all he had to do was to utter the code word and Harry would be in a hypnotic trance. This Mr. Jochems accomplished successfully. He was discreet enough not to ask me why I desired such a quick method of hypnotizing Harry. However, I told him that I wanted to do some telepathy experiments with Harry under hypnosis, and wanted to be able to hypnotize him when he least expected it. We did in fact carry out such hypnosis-telepathy experiments.

After Mr. Jochems had hypnotized Harry by the code word, which in this case was "Blue Geranium," I then asked him to do an ordinary age regression of Harry year by year, starting from the present back to the year of his birth, and probe for any unusual language experiences that Harry might have had. Mr. Jochems carried out such an age regression, and although he did not probe specifically for Egyp-

118   THE SACRED MUSHROOM

tian experiences, he did not in the course of his regression turn up useful evidence of such experience.

Then I asked Mr. Jochems to regress Harry beyond his birth, and asked to be present at this hypnosis. This was the first hypnosis for age regression in which I had been present. On September 25, 1956, Mr. Jochems tried such a regression. In the course of this regression Harry described a previous "lifetime" as a German living in the middle of the nineteenth century, but no unusual details turned up in the description of this "lifetime."

After this session, I finally revealed to Mr. Jochems what my real purpose was in these age-regression experiments. I informed him that I wanted to ascertain whether Harry had had in his lifetime any experiences that might explain the written Egyptian material which he had produced. He now agreed to probe specifically for Egyptian experiences. Mr. Jochems again regressed Harry Stone year by year through his life but did not turn up any evidence significant of experience with things Egyptian, other than art school and museum encounters with statuary. He then regressed him before his birth and again encountered the German "lifetime" in the middle of the nineteenth century. Then he regressed him by telling Harry by strong suggestion to go back as far as he could into his memory. I was present at this interrogation. I will now give an extract of this hypnosis experiment of the first of October, 1956.

"The time is 11:35 P.M. Mr. Jochems is the hypnotist. Harry Stone is the subject, and Dr. Puharich is present. Harry is immediately hypnotized with the code word 'Blue Geranium.' He is regressed to the age of three and speaks in Dutch baby talk. He is then told that he is now in the womb. Harry curls up in a fetal position and looks blissfully happy. Mr. Jochems then regresses him backwards in time with the count of ten.

Mr. Jochems says, 'You are going back to many lives ago, you feel wonderful. You can really say what you see. What is your name? Speak your native language.'

Harry did not reply to any of these suggestions.

Jochems: 'What do you see?'

There is a long pause as Harry does not reply.

Jochems: 'What would you like to tell me?'

At this point Harry points to his ear with the left forefinger and then makes his first vocal statement, which is: 'PAR UP KA.' Harry then holds his right hand up to his chin in the 'promise' sign, and

THE SACRED MUSHROOM 119

then makes the 'secret' sign with the first two fingers of his right hand over his lips.

Mr. Jochems: 'Who would you like to speak to in secret?'

Harry then lies down on his back and makes the following hand sign language: 1. Cups both hands over his abdomen. 2. With his left hand he places his fingers in his mouth and opens his mouth. 3. With both hands he makes the sign of a flying bird over his face. 4. Both his wrists are crossed over the abdomen. 5. He grasps the right thigh. 6. Brings his left hand to his mouth again. 7. The left hand is brought to the left jugular region. 8. The right thumb and forefinger are placed on his sternum. All this was silently carried out, and then Harry indicated that he wanted a pencil and paper, and these were handed to him. He then drew six Egyptian hieroglyphic characters which were three of the semicircular Tep signs and three of the Nefert signs which he had drawn a number of times before. He then relaxed and lay back, apparently in deep trance. Without any suggestion from the hypnotist he then woke up. The hypnotist re-enforced his waking up by positive suggestion. This was the end of the hypnosis of Harry Stone."

This hypnotic session was very interesting. By the use of the regression technique we were not able to uncover any conscious or unconscious experience of Harry with significant Egyptian material in his own lifetime. When regressed in a number of sessions to so-called "former lifetimes," nothing very interesting turned up in the way of Egyptian knowledge or associations. Now in this last session recorded there is the only presence of Egyptian knowledge out of six hypnotic experiments conducted. These can be interpreted for the reader. When Harry pointed to his left ear, we assume he meant "listen." The phrase PAR UP KA is an Egyptian phrase, and it can be translated as "[Ascend], the doors of the soul are opened." This statement was followed by the sign language described, which is actually a part of the silent-mushroom ritual which Harry had performed in trance a number of times before. The hieroglyphs written at the end of this hypnotic session do have a meaning and can be translated as: "Good! [In the moral sense]."

The entire message, that is, the vocalized Egyptian, the sign language, and the hieroglyphs are obscure in their meaning. It is interesting that under interrogation during this demonstration, Harry was very uncommunicative and would not respond to questions directed as to who he was and what it was that he was seeing. The real ques-

120 THE SACRED MUSHROOM

tion comes up here as to whether Harry may not have slipped from the hypnotic trance state directly into the deep trance state which he had exhibited spontaneously many times before. There is no way of answering this question, since the two states, that is, the induced hypnotic trance and the spontaneous deep trance, are really indistinguishable. The only criterion that one has in differentiating between the two is what the subject says. In this case what he said did not greatly clarify the problem.

My conclusion from this elaborately prepared hypnosis experiment was that we did not turn up any evidence within Harry Stone's own experience to account for the Egyptian vocalized statements, the sign language, or the written hieroglyphic statements. We could go no further in this type of deep exploration. The presence of the few enigmatic Egyptian phrases which his last hypnotic session had yielded was difficult to interpret. It is my personal opinion that this, too, represents a phenomenon similar to his spontaneous trance states and cannot be easily accounted for on the basis of his own experience in this lifetime. Three years of close observation and experience with Harry Stone had not given me any opportunity to build up a case which could prove that he was in any way fabricating, either consciously or unconsciously, the material which he delivered in trance. I could come to only one conclusion. Harry Stone had exhibited a true spontaneous, deep-trance phenomenon which had begun on or about June 16, 1954, and which had ended on February 7, 1956. The beginning of the trance exhibition was spontaneous and its ending was spontaneous. This was perhaps the most unusual case I had ever observed of spontaneous deep-trance phenomenon.

The unanswered question was, who was responsible for the intelligence which Harry Stone had delivered?

Having satisfactorily settled the question of Harry Stone's motives in this series of writings; and, furthermore, since the entire story was buttressed by such external evidence as Mr. Wasson's discovery of a mushroom cult in Mexico, and the finding of the golden mushroom in Maine, I now had to look at the internal evidence presented by the corpus of Egyptian material which had been delivered in the trance state. This material was consistent from beginning to end for the archaic character of the written language. The alleged Ra Ho Tep personality was traced, and a historical Ra Ho Tep was found whose lifetime coincided with that of the archaic period of the language which Harry had transmitted. The Ra Ho Tep personality intelligence when translated formed a coherent body of beliefs which can also be related to the period of the historical Ra Ho Tep. The evidence for this is found largely in the written characteristics and ideas present in the Pyramid Texts which were written during the period of the historical Ra Ho Tep and the hundred succeeding years. The evaluation of the internal evidence by this corpus of writing by the entranced Harry Stone was my next detective job.

chapter X

The Egyptian writing had begun on June 16, 1954, and the last legible writing was produced on September 8, 1955. Although Harry had gone into trance four more times after September 8, the products of these sessions were unintelligible. The only clear information that appeared in these four trance sessions was the sign language. This encompasses the entire span of this phenomenon as it concerns the Egyptian language and Harry Stone.

During this period Harry delivered forty-nine separate written hieroglyphic messages. In this written material one hundred and eighteen different Egyptian hieroglyphs were used. Gardiner[1] lists about seven hundred and fifty hieroglyphs as used during the Middle Kingdom period of Egyptian history. Therefore, Harry's production of hieroglyphs would constitute about a sixth of this list of Egyptian hieroglyphs. Now of the hundred and eighteen different hieroglyphs used in the forty-nine separate written messages, ninety-two are of a familiar form from the Old Kingdom period on. The remainder of the hieroglyphs (twenty-six) can be classed in two groups. Sixteen of the twenty-six hieroglyphs are of an archaic form which can be dated from the Old Kingdom and which dropped out of use by the XIIth Dynasty or so. These, therefore, point to a date before this time. The ten remaining hieroglyphs are problematical as to their phonetic value and their meaning. Some of these could be loosely interpreted as pictographs by comparison to similar hieroglyphs; or they could be interpreted sometimes on the basis of the context in which they ap-

[1] *Egyptian Grammar*, Sir Alan Gardiner. Oxford University Press, London, 1950. Budge states that a total of 2860 hieroglyphs are known, including variants.

124   THE SACRED MUSHROOM

peared, or on the basis of additional information supplied by Harry, either in spoken English or in vocalized ancient Egyptian. Nevertheless, these ten hieroglyphs are not definitely translatable.

In addition to the written material, Harry produced thirty-six vocalized Egyptian phrases. No one knows what the ancient Egyptian language sounded like. Scholars have been able to reconstruct what they think was the sound of the language, but such reconstruction is problematical. These utterances presented the greatest difficulty in translation because they were pronounced in syllabic form, and hence the beginning or end of a word was never made clear. Since in the ancient Egyptian language many different hieroglyphs have the same phonetic value but have different meanings according to the figure represented by the picture, it becomes very difficult to assign a definite hieroglyphic meaning to any given phonetic value. The translations given for the vocalized Egyptian are presented with the utmost reserve for this reason.

The biographical data furnished by the alleged Ra Ho Tep personality during these trance sessions was very meager. Ra Ho Tep identified himself as a King, or as a royal acquaintance. He gave the name of his wife as being Nefert. When asked what place he came from, he uttered the word MEDU. It is assumed that Medu is the same as the known place in Egypt called Medum. The Ra Ho Tep personality associated Medu with Snefru when he used the phrase MEDU MA SNEFRU. This means "Medum together with Snefru." Since it is generally assumed that one of the Pharaoh Snefru's monuments is at Medum, I have interpreted the word MEDU as meaning what we now call Medum. In addition to these brief identifying phrases Ra Ho Tep has given the name Ptah Khufu. It is known that Ptah Khufu was the Pharaoh who succeeded shortly after the Pharaoh named Snefru. Therefore, I concluded that in giving these two names, Ra Ho Tep is identifying himself with the time and place of these individuals. If Khufu and Ra Ho Tep are indeed sons of Snefru, then they are either brothers, half brothers, or stepbrothers.

Assuming, therefore, that the Ra Ho Tep personality of the trance utterances of Harry Stone is to be associated with these two names and their period, a historical search was initiated. This turned up the work[2] by Sir W. M. Flinders Petrie, describing in detail the burial tomb of the historical Ra Ho Tep at Medum. The historical Ra

---

[2] *Medum*, Sir W. M. Flinders Petrie. David Nutt, London, 1892.

THE SACRED MUSHROOM   125

Ho Tep was buried close to the pyramid assigned to Snefru. Ra Ho Tep seems to have been buried at about the same time as that assigned to this Pharaoh. The personal information derived from his tomb inscriptions is again very meager. Here we find that Ra Ho Tep is listed as the "King's Son, from his own body." We find that in Snefru's list of children there is the name of the Prince Ra Ho Tep. On the basis of such information, historians have assumed that the Ra Ho Tep buried in the Medum cemetery is the son of the King Snefru.

In the title list of the historical Ra Ho Tep, Petrie reports that fourteen titles are assigned to him. Of the fourteen titles only three are translatable with any precise meaning. The other eleven titles are so ancient that they were dropped from use shortly after the IVth Dynasty and their meaning has not yet been reconstructed. Of the three titles that are understandable, the first lists the historical Ra Ho Tep as the high priest of Heliopolis. During this period of Egyptian history, Heliopolis was the equivalent of the modern Rome as a religious center. From Heliopolis, which was called Anu in its day, spread the main religious doctrines characteristic of the Old Kingdom of Egyptian history. We shall find that some of these doctrines, known principally through the Pyramid Texts, are reflected in the writings of the Ra Ho Tep personality, as manifested through Harry Stone. The other two titles of Ra Ho Tep listed in Petrie's book are: "Member of the Southern Tens," and "Captain of the Host."

As high priest at Heliopolis the historical Ra Ho Tep was certainly one of the leading men in the Egyptian hierarchy, and "Captain of the Host" is a title that might apply to several military grades, from general to lower ranks, and it is suspected by historians that Ra Ho Tep was really the commander in chief. In the scenes from the tomb, we see the King's son Ra Ho Tep accompanied by the Lady Nefert, and the latter is designated as "a royal acquaintance." Upon the stela of the outer court is an inscription which probably is to be read, "the royal acquaintance, the King's son of his body, who has attained the reward of merit, BU-NEFER." A rare peculiarity is that he should be called "a royal acquaintance" as well as "the King's son."[8]

With the meager personal data from the statements of the Ra Ho Tep personality, and from the historical Ra Ho Tep's tomb, we now try to ascertain if there is any relationship between these two figures.

[8] *Ibid.,* p. 37.

126   THE SACRED MUSHROOM

I have closely examined all the scenes reproduced by Petrie from the historical Ra Ho Tep's tomb; and I have minutely studied all the utterances and writings of the Ra Ho Tep personality. In comparing the two, I find that the Ra Ho Tep personality has not used any of the titles which are found in the tomb of Ra Ho Tep. This presents us with a rather interesting problem. If Harry Stone were consciously fabricating the Ra Ho Tep personality, and if his source of information was from this book by Flinders Petrie, which is one of the two works extant on the subject, then he surely would have used some of the material from one of these two books in order to establish his identity. The fact that he has not leads me to believe that Harry did not at any time in his life learn the material from the Ra Ho Tep of either of these two works. If he had, and was using such material consciously or unconsciously, it surely would have appeared somewhere in his utterances. On the other hand, if none of the material identifying the historical Ra Ho Tep is present in the writings and statements of the Ra Ho Tep personality, then the question is raised as to whether there is any relationship between these two figures.

We have the following facts to consider. The Ra Ho Tep personality does identify itself, both by writing and by the few allusions to names and place, with the period of the IVth Dynasty. The titles given by the Ra Ho Tep personality are not those which are found in the tomb of Ra Ho Tep. The only connection between the Ra Ho Tep personality titles and those in the tomb is the fact that they are priestly titles. Since the historical Ra Ho Tep is listed in this tomb as the high priest of Heliopolis, one might assume that these priestly titles given by the Ra Ho Tep personality only serve to define further the specific functions of the priest. But this is only an assumption.

The Ra Ho Tep personality has given us a number of details of his life. But these are highly personal details and cannot be confirmed by the meager details found in the Ra Ho Tep tomb. Therefore, there is little possibility of matching the details of the Ra Ho Tep personality with those of historical Ra Ho Tep. We must consider the remote possibility that the name Ra Ho Tep as it appears through Harry Stone is a fictitious name. By this I mean that the phrase Ra Ho Tep, as used throughout ancient Egyptian history, also means "a royal acquaintance." If the Ra Ho Tep personality is merely identifying himself as "a royal acquaintance," we must ask the question, who is the royalty referred to? Now the Ra Ho Tep personality pre-

THE SACRED MUSHROOM 127

fers to identify himself more frequently by the appellation, En Katu. There is a title in the Ra Ho Tep tomb which can be pronounced En Katu. The interesting point here is that the Ra Ho Tep personality has not used the hieroglyphic spelling which is found in the tomb. Instead, he has constructed the equivalent phonetic value with an entirely different set of hieroglyphs. The hieroglyphs used by the Ra Ho Tep personality in this respect are unique in their make-up. However, this is a small clue that is not sufficient to connect the two Ra Ho Teps.

From the meager data available thus far we are not justified in assuming that the Ra Ho Tep personality is to be identified with the historical Ra Ho Tep. As with much of this trance material which we have been considering, a solution to this problem has not been reached. In addition to the personalities already mentioned, the Ra Ho Tep personality brings up names like: Amenenhotep, Anahatet and Antinea, Nakita, and Ptah Khufu. None of these names can be fitted into a historical context with the exception of Khufu. However, it is well to remember that the period of the historical Ra Ho Tep is very sparse in historical detail. We have before us primarily the great historical monuments associated with the Pyramid Age, and only a very small amount of historical data. Perhaps, in time, the identification of these other names can be accomplished.

We now take up the Egyptian writings made by Harry Stone in the trance state that give the titles which the Ra Ho Tep personality has associated with himself as hallmarks of identification. The first one was made on June 16, 1954, and is written as "Ra Ho Tep, the King. Nefert [his Queen]." This phrase can also be translated as "A royal acquaintance, Nefert." The next title is similar: "Ra Ho Tep, it is my name. En Katu, life." In giving this secondary name, En Katu, there was used for the first time a scepter which is the totem mark of the Oxyrhynchite nome, an area geographically close to Medum.

On August 24, 1954, there was written: "Ra Ho Tep, the ruler, in his name of the physician." It is interesting that in this hieroglyphic statement the Ra Ho Tep personality clearly identifies himself as the ruler, or Pharaoh. The written material from the tomb of Ra Ho Tep does not clearly designate him as a King; he is only listed as a prince, a son of the King, and a royal acquaintance. We, too, encounter the oddity that the writing gives both the name Ra Ho Tep and the title, a royal acquaintance.

128   THE SACRED MUSHROOM

On August 29, 1954, there was again written: "Ra Ho Tep, my name." The Ra Ho Tep personality spends a great deal of time talking about Antinea, whom he identifies with Alice, and it is interesting that she is also delineated as "a royal acquaintance," in one of the writings. On August 29, 1954, there was written: "Ra Ho Tep, the adored King," or, "a friend of the adored King."

On August 31, 1954, there was written: "Ra Ho Tep, Ruler of the North and South, United." The symbol of Kingship used up to this writing by the Ra Ho Tep personality was always the Sut emblem of lower Egypt, or the North. This is the only instance in which he identifies himself with the United North and South of Egypt.

On September 4, 1954, we have the vocalized statement: "The lord of life enfolds [in protection] Ptah Khufu, and Antinea." Now, as stated many times before, the Ra Ho Tep personality definitely associates the name Antinea with Alice. However, he has never clearly identified the name Ptah Khufu with any individual. Since at the time of this statement there was present in the room only Harry, myself, and Alice, one wonders whether this name could refer to Harry or me. But there was nothing in the statement which established an association of this name with either of us.

On October 5, 1954, there was written "Tehuti, the creative one. Tekh." In this message the name of Tehuti is also given as Tekh. Tekh was another of the known names given to the god Tehuti. (See Appendix 2, DRAWING No. 7, for full details.)

The preceding titles for the Ra Ho Tep personality, and the identification of the Tehuti personality, mark the two main characters that manifest through the entranced Harry Stone. These names and titles have all been given as the English translations of archaic forms of the vocalized and written ancient Egyptian language. Because of the archaic character of this form of expression, the translations are offered with the utmost reserve.

chapter XI

We will now consider the Ra Ho Tep personality statements which can be interpreted as referring to a sacred-mushroom rite. The first indication that one of the main purposes of these messages in the ancient Egyptian language was to inform us about a long-lost cult of the mushroom occurred on June 16, 1954. At this time there was made a drawing of a mushroom, and a verbal description was given of its color and other characteristics. But at this time the only information given was that this mushroom could "take a man out of his body." It was quite a stretch of the imagination, at this time, for me to infer that this statement also meant the phenomenon of fissioning the soul, or consciousness, from the body. However, this inference has been amply borne out by historical evidence and tradition, particularly from eastern Siberia.

On June 16, 1954, there was written: "Anubis [as AMA UT]. He who guards the secrets of plants out of which Ra [sun god] appears." Now this is the first of a number of references which are given to a god who assumed the form of a dog and who was revered throughout ancient Egyptian history as the god of the underworld, Anubis. He is traditionally portrayed particularly as the "one who guards over secrets," and these apparently refer to secrets connected with the other world. Anubis is also present in the Judgment Hall scene of the Book of the Dead in connection with Tehuti, when the heart or the life actions of the deceased are weighed in the great balance.

The plant and the secrets guarded over by Anubis were progressively revealed in a series of hieroglyphic statements. The technical details of this translation are given in Appendix 2, and I present here

129

130 THE SACRED MUSHROOM

only the free translations of these writings. They are not presented in the chronological order in which they were recorded but rather in an order that reveals their internal consistency. The number before each line refers to the number of the drawings as listed in the Appendix.

1. Ra Ho Tep is my name. The red crown plant of ascension [over] life.
2. A friend returns [with] the red crown of ascension over life.
3. Great offering [of the] red crown plant [is] coming into being.
4. Exalted growing place! A friend of the King, double son of Ra.
5. Double offering [of the] golden plant.
6. Ra Ho Tep, the adored King offers the life-giving plant to two sons. Or this may be read: The adored friend of the King offers the life-giving plant to two sons.
7. A friend of the King, Tekh [dispenser of life]. Tehuti, creator of the Aakhut Pyramid.
8. The supernatural power of [a pair] of AAKHUT [mushrooms].
9. The divine journey is offered you by the great gods of life to the end of eternity.

There is a clear and consistent description here of a plant, either as gold or red in color, and its sacred nature. In passage 9 in the phrase, great gods, there is a hint that the gods may be the Ennead, or the great nine ruling gods of Egypt. This hint is more clearly spelled out in a hieroglyphic passage of October 5, 1954. In translation: "A bond is broken [released?]. The secret is broken [released?]. Hear the Ennead speak of the alabaster vessel of life." At the time this statement was recorded it was quite impossible to translate it and get any clear meaning from it. This could only be done after the entire story was complete. I interpret this statement to mean that the time had come to reveal or divulge a long-standing secret, presumably in connection with the sacred mushroom. Nowadays, we would say that a secret has been declassified.

The reference to the Ennead is in itself interesting. Heliopolis was the center of a religious system which had for its pantheon nine great gods, and these have been called the Ennead, which means the Nine. The Nine of Heliopolis are Atum, Shu and Tefnut, Geb and Nut, Osiris and Isis, and Seth and Nephthys.

The Ennead were believed to be the all-powerful ruling gods of the world, and their chief spokesman was to be found in the high

THE SACRED MUSHROOM   131

priest of Heliopolis. If a Ra Ho Tep personality did live during this period, he would affect the religious beliefs of his own time, even though he is speaking over four thousand years later. This reference to the Ennead should be considered as one of the bits of internal evidence which lends coherence and consistency to a Ra Ho Tep whose date is assigned to the IVth Dynasty. And we must take it into consideration in our total evaluation of the Ra Ho Tep phenomenon.

The passage given under number 8 above was written in hieroglyphs on September 4, 1954. This was the first clear message that pointed to the reason for the Ra Ho Tep personality manifestation. This was the prelude to a gradual unfolding of mysteries connected with a magic AAKHUT, a golden bough, or ladder, known in ancient times. It is interesting that the plant in question, which I assume to be a mushroom, is shown in hieroglyphic characters with one of the signs represented as a ladder. In accordance with the strange mode of expression of ancient Egypt, I interpret this to mean that it is the plant whereby ascent to heaven is effected. The ambiguity as to whether it should be called the "golden ladder," or the "golden bough" arises from the form of presentation of the hieroglyphs.

It is interesting that much later in history, throughout the Classical Age of Greece and Rome, there existed a legend of the golden bough. This has been voluminously described and analyzed by Sir James Frazer in his monumental work, *The Golden Bough*. Here the golden bough is assumed to be a talisman, by whose presentation and power one gained access to the other world, or the underworld. We find the meaning of this golden bough clearly described in *The Aeneid*, where the hero, Aeneas, in order to gain entrance to the underworld, and to be able to return to the world of the living, has to find and present to the ferryman the golden bough. (See Appendix 3, page 193.)

On the same date, September 4, 1954, Harry Stone, in trance, made the following statement in English in answer to my question as to how the mushroom was used. He said, "By opening the door, by stepping in, and leaving." Again this statement is an enigma wrapped in obscurity and its meaning was only clarified by subsequent sign-language demonstrations by the Ra Ho Tep personality. This statement apparently refers to the great door which the ancient Egyptians imagined to guard the entrance into the other world, the Tuat, or Restau, as they called it. And it is precisely in this context

132 THE SACRED MUSHROOM

that we find the classic writers describing the golden bough. It, too, must be presented to the gatekeeper, or the ferryman of the other world.

On August 29, 1954, we have a series of pictograms which can only be interpreted as an ideogram: "Salt for the tongue and head. Kenu [the god of the sun boat]. KHEPER [the sign of coming into being]," is a drawing of the sun boat with a KHEPER emblem on the sail. This statement was also an enigma at this time, and it was only in subsequent demonstrations by the entranced Harry that its meaning became clear. "Salt for the tongue and the head," means precisely what it says. Salt is to be applied to the tongue and to the top of the head as part of the sacred-mushroom rite. This was demonstrated by Harry after we had found the golden mushroom. The Kenu, which was written in hieroglyphs, was personalized by drawing a cartouche or an encircling line around the name. This would make it the name of the god of the sun boat, or he who directed the sun boat. It must be remembered that the sun boat had two distinct meanings to the ancient Egyptians. The first is that the sun, personified as Ra, made the journey across the sky every day in a boat. The second meaning is that one ascended into heaven, or the other world, by means of the sun boat directed by the god Henu, or Kenu.

The KHEPER is symbolized by the beetle. Since the beginning of ancient Egyptian history the beetle was conceived of as a symbol of self-begotten creativity. They probably derived this symbol from watching the beetle lay its eggs in the mud and roll them up into a ball. The ball was left in the hot sun and the eggs would hatch. This was the great symbol then, of creation, or of coming into new being. The picture of a sun boat with the beetle symbol prominently displayed on the sail undoubtedly refers to the role of the sun boat in carrying the deceased into a new state of being in the afterlife, and thus symbolized the "coming into a new being."

On October 5, 1954, there was written: "The golden oracular headband [plant] through which Ra appears." This statement too, is obscure, unless one becomes acquainted with the thinking of the ancients. It is known, for example, from *The Aeneid* that the priestess or oracle who divined future events assumed a gold headband as part of the ritual of prophecy. There are other allusions to a gold headband as one of the props of prophecy in the writings of ancient times, including those of the Egyptians. Hence we may assume that the gold headband here referred to is the usual headband donned for the purpose of uttering prophetic statements. The remarkable part about this statement is the fact that it is the sun god Ra as a plant who will manifest or speak through the gold headband. Now this may also be taken as a figure of speech in which the "gold headband" refers to the golden mushroom. We have had another statement in which there is reference to the appearance of Ra through plants, particularly a golden plant. Hence this statement can be grouped with other statements that refer to the divinatory powers allegedly conferred by a golden mushroom in the beliefs of the ancient Egyptians.

On June 20, 1955, we have two separate vocalized Egyptian statements by the entranced Harry. The first is: "KA HB NOU MI." This is translated as: "The KA [soul] of Tehuti looks on." The statement following this is a single word given as SHU. These two statements must be considered together. I must first describe the context in which they occurred. Harry was giving one of his silent sign-language rituals. He uttered the word SHU at that point in the ceremony at which I believe the sacred mushroom is given to the initiate. Now the statement by itself again leaves us baffled unless we can get into the framework of ancient Egyptian thinking.

By referring to the Pyramid Texts we find in line number 1377a: "Tehuti who is in the SHU of his bush, put the deceased upon the tip of thy wing." Now the word SHU is usually translated as shade, or "the shade of his bush." Now what does the word SHU really mean? SHU refers to the air god who is one of the great Ennead. The god Sept is also known as SHU. And one of the words for the sun, besides Ra, is SHU; so we begin our inquiry with three possible meanings for SHU, all of them referring to ancient gods. We have the fourth meaning in this sentence quoted from the Pyramid Texts which is given by the translator as "shade." It is interesting that this word for *shade* is made up of three hieroglyphs,[1] and the second one, which is the determinative, is in the shape of a mushroom. This is a very controversial point in the interpretation of Egyptian hieroglyphs, that is, what does this mushroom-shaped figure stand for? Most Egyptologists take the point of view that this stands

[1] Kurt Sethe. *Die Altaegyptischen Pyramidentexte*, Leipzig, 1910. Band II, L. 1377a.

134   THE SACRED MUSHROOM

for an umbrella or the shade which was held over the head of the Pharaoh by an attendant.

However, if we look at it in the particular context which I have cited, we see that it may have other meanings. It does not make much sense to say that "Tehuti, who is in the shade of his bush" puts the deceased on the tip of his wing and then takes him up to heaven. It would make more sense if we accepted the Mexican interpretation of the meaning of the mushroom, and assumed that it is Tehuti who is actually in the mushroom, in the way that the Mexicans speak of their god *being the mushroom.* In other words, when a Mexican *curandero* uses the mushroom, it is no longer he who is speaking, but it is the mushroom who speaks, and the mushroom is personified as a god.

The sentence from the Pyramid Texts gives us further enlightenment because the phrase, "the shade of his bush," can have other meanings. It is well known that the mushroom, which we have identified, namely the *Amanita muscaria,* usually grows in association with certain trees. This may be one of the reasons that trees were worshiped in ancient times—not only because they were the source of fuel and heat and material with which to build, but because they were the source of the life of the sacred mushroom. There are many primitive and ancient cultures which worship trees and assume that a spirit resides in the tree. This sort of personification of trees would make much more sense if we assumed that the ancients believed that the soul, or god, of the mushroom resided in the tree until its emergence from the ground during certain seasons of the year. For these and other reasons which I will go into in the next chapter I feel that it is necessary to re-evaluate the meaning of the ancient Egyptian hieroglyph which has the shape of a mushroom.

On June 23, 1955, there was written in hieroglyphs: United. On June 24, 1955, there was written in hieroglyphs: "Hasten to be purified. You are united at the threshold of the Door of Eternity." At the same time Harry stated in English that he saw in an imaginative situation four people standing before a huge door. He recognized of course that the door was purely symbolical. The four people standing before the door were Harry and myself in contemporary dress, and Alice and Betty, as Antinea and Nakita, respectively, in ancient dress. Harry had the clear impression that we had all assembled in front of this door for some specific purpose, but that we did not have the key to get through. I cite this episode in some detail in

THE SACRED MUSHROOM   135

order to clarify the meaning of the Egyptian statements, and the events that followed.

On June 30 there was written the longest single statement made by the Ra Ho Tep personality which I have already translated as follows: "This is the name of the place of waters from heaven," etc. (See page 71.) Wainwright[2] states that in the archaic HEB-SED ceremony, a pair of "shades" represent the gods MIN and AMUN. By the time of the Vth Dynasty the pair of shades is replaced by a sign that means "waters of heaven." Thus the Ra Ho Tep personality could be using the phrase "waters from heaven" to hint that a "pair of shades" is coming into being.

On July 1, 1955, there was vocalized the following Egyptian phrase: "ASH NUAH." Nuah is a god mentioned only once in the entire corpus of the Egyptian literature, and this is in the Pyramid Texts. Nuah is a dog god who pulls the boat of the deceased on his way to heaven. Interestingly enough the deceased is called the "great Nuah" in the Pyramid Texts. The Texts show that when he is called the great Nuah he is also at this stage of the funerary ceremony purified by water, and then begins his ascent to heaven. We can therefore relate this reference to the god Nuah to the previous litany of purification by water. The word ASH[8] is also interesting in that it refers to a dog god whose functions are unknown. It may be that we have a duplication of gods here, that is, one being Ash, and the other being Nuah, both of them being dog gods, and the latter is clearly identified with a divine journey to heaven.

On July 3, 1955, we have the vocalized Egyptian phrase translated as: "The great Sphinx is in our midst." The statement that follows is: "The great water from heaven enters a golden plant." The exact meaning of having in our midst the Sphinx is obscure, but in the second message we again have a reference to the great water from heaven, and this time it is supposed to enter a golden plant.

[2] G. A. Wainwright. *The Sky Religion of Egypt*. Cambridge, 1939, p. 21. The archaic scenes show two fans or shades in the HEB-SED ceremony:

These are drawn either as sunshades, fans, or lotuses. By the Vth Dynasty the above symbol becomes: "waters of heaven."

[8] *Ibid.*, p. 85. The god Ash is also believed to be one of the central figures in the HEB-SED ceremony.

136 THE SACRED MUSHROOM

I have already recounted at length the statements given under hypnosis by Alice Bouverie on July 4, 1955. The reader will remember that this message indicated rather clearly where on the grounds a certain kind of mushroom could be found. Such a mushroom was found on July 6, 1955, and it was golden, and it was identified as the *Amanita muscaria.*

When we consider the slow unfolding of information given in the ancient Egyptian language about a secret and sacred-mushroom cult over a period extending from June 16, 1954, to July 3, 1955, with the finding of a golden mushroom on July 6, 1955 we have before us a unique drama. We have wrapped in the obscurity of a dead language certain arcane hints and specific information. We have from another source intelligence which leads to finding the kind of mushroom which has been previously advertised in an unusual manner in the ancient Egyptian language. After the mushroom was in our possession experiments revealed that it does indeed have remarkable properties, i.e., anesthesia, inebriation, etc. Thus the entire story, both from the internal evidence and the external evidence, leads us to place some credence on the ancient Egyptian utterances and writings of the Ra Ho Tep personality.

Now the third part of the drama unfolds after we have found the mushroom. This material has an anti-climactic tone when compared with the writings we have just analyzed. On July 19, 1955, while doing a routine telepathy experiment Harry stopped his work but did not go into trance. He was stopped by a vision which was present only in his mind. He saw an enormous human figure standing over the ocean with a cord of rope hanging from each arm, and holding up the fiery ball of the sun. This vision held Harry's attention for fifteen minutes. When I asked him what it meant, he could only say, "Don't let the sun fall into the water!" This was the only meaning he could give to this extraordinary vision. I myself cannot interpret it.

On July 27, 1955, the Ra Ho Tep personality vocalized in ancient Egyptian the following statement: "We are filled with the fragrance of brotherly friendship." On the same date: "Ra in his name of eternity. Enter the door of the twofold strength." On the same date: "The great Sphinx is in our midst. Become like the ruler Ptah Khufu. Do like the King of old!"

The last statement must be explained in the light of the circumstances surrounding its utterance. This was one of the spontaneous

THE SACRED MUSHROOM   137

trances of Harry Stone which occurred about 11:00 P.M., after he had gone to bed. I was called from the laboratory and informed that Harry was behaving rather peculiarly. I went to his bedroom and found him in an agitated trance state. During the course of this trance he uttered the statements which I have given. When he came to the statement, "Become like the ruler Ptah Khufu," he stared at me with his piercing blue eyes for many minutes before he uttered this statement, and made it most emphatically. I could not help feeling that it was addressed to me in a highly personal way. The next statement, "Do like the King of old," was again directed to me, and repeated most emphatically three times; I could not escape the feeling that I was being charged with a highly personal commission, but since the statement was so fragmentary I knew not what.

On August 7, 1955, Harry, as the Ra Ho Tep personality, demanded that the golden mushroom be brought to him. I have already described this episode in detail. It was important for two reasons. First, the Ra Ho Tep personality enacted the ritual use of the mushroom. This showed that it was placed on the tongue, and rubbed into the scalp at the site known as the Aperture of Brahma. Second, it showed the inebriating power of the mushroom, and the increase in telepathic acuity of Harry Stone while under its influence.

These two demonstrations confirmed the hints of a sacred-mushroom rite and the alleged "divinatory" powers of certain mushrooms, in this case, *Amanita muscaria.*

This dramatic episode seemed to be the climax of the manifestation of the Ra Ho Tep personality. This can be partly seen by a close study of the Egyptian statements which I have already given. And it is further emphasized by the remaining Egyptian utterances of the Ra Ho Tep personality. On September 8, 1955, Harry was at the seashore fishing when he went into uncontrollable deep trance. I was called by Betty and arrived on the scene ten minutes after trance had started. By the time I got there Harry was beginning to come out of trance. But Harry in the meantime had found a piece of charcoal on the beach and inscribed on the surface of a rock four Egyptian characters which I translate: "Eternity is watching."

This was the last significant statement that Harry uttered in the self-induced trance state. One other statement appeared a year later when we were doing hypnosis experiments with Harry. At this time he made the statement: "The doors are opened for the soul." He

THE SACRED MUSHROOM

finished this hypnotized trance session with a hieroglyphic statement which means: "All is well."

The few remaining trances which occurred definitely showed a weakening of the grip of the trance phenomenon on Harry. He attempted several times to write hieroglyphs, but the statements were practically illegible. The same holds true for a few utterances he made in the Egyptian language. On February 7, 1956, he had his last feeble trance, and at this time he was not able to communicate anything clearly. He has been free of this condition up to the present time.

He and I have discussed this strange episode in his life many times since then. He does not understand it now any more than he did when it was going on. He is thankful it is all over, and that he does not have to worry about going into trance in situations where it might be embarrassing to him personally, or even dangerous. I was grateful that this experience had come across our paths, but thankful, too, that it was over. The study of these strange messages in the ancient Egyptian language gave me a unique opportunity to explore a rare aspect of the mind. I now ask the reader to enter a long labyrinth with me as we go back into ancient Egyptian ideas and beliefs.

chapter XII

THE word "mushroom" is conspicuous by its absence from the ancient Egyptian language. This either means that they did not know the mushroom, and this is unlikely, or that they held it in such high reverence that it was the best-kept secret of their culture. The only reason that we have for the latter supposition comes from the Ra Ho Tep personality statements. This can hardly be considered as evidence unless it can be found to fit in naturally among the historical writings and beliefs of the ancient Egyptians.

The clues from the Ra Ho Tep personality source are centered around a word presumably associated with the mushroom, vocalized as AK KHUT, AAK KHUT, or ANGKHUT. In the hieroglyphs this appears as:

The vocalized word could have the same meaning as the Egyptian phrase for the tree of life, KHUT PU ANKH. If this were true then we would rest the case here and not seek for a definitive word for the mushroom. The texts do not tell us what this "tree of life" is, and we are entitled to continue the search for the meaning of this phrase.

But the hieroglyphs as a pictorial representation of the sound AAK KHUT are worth more than a thousand words, to paraphrase the Chinese saying. The AAK part of this word, is the word for a

140 THE SACRED MUSHROOM

ladder, which is represented by the third hieroglyph. The second syllable KHUT is the word for a tree, wood, or plant, and corresponds to the first hieroglyph. The middle hieroglyph is unknown in the Egyptian language, and in this case we cannot assign any phonetic value to it, and assume that it is a determinative which represents the object AAK KHUT.

The middle hieroglyph commands our attention first because it may contain the meaning of the entire group. There are known three similar hieroglyphs to which we may compare our unknown hieroglyph. (See Appendix 2, DRAWING No. 8, pages 187–88.) The first is a sign which means "gold" or "golden." This has an upright loop, which represents an unknown object, and a mouth sign, RA, across it. The mouth sign of course represents the sun god Ra, and this may be the source of the meaning gold from the golden color of the sun.

A second sign which approximates our unknown sign means to "guard" (something); in some examples to guard Egypt. Here we have the idea of protection or secrecy to consider.

A third sign which approximates our unknown sign is a representation of the phallus within the vulva. As such it represents the idea of the generative function.

Each of these three known signs from ancient Egyptian sources that approximates our unknown sign contains some of the metaphors associated with the mushroom in other cultures. The sign for gold is obvious because of the golden color of our mushroom. The sign for guard can be associated with the secret and sacred nature of the mushroom cult as, for example, known in Mexico. The third sign, that of the generative function, as symbolized by the male and female organs, is a well-known association for the mushroom, and Wasson has described in great detail the origins of this symbolism. It can also represent the god Min who is intimately associated with the archaic HEB-SED ceremony. While these three symbols fit into the pattern of mushroomic lore, they do not necessarily tell us that the ancient Egyptians attached the same meaning to this unknown symbol, since we have not yet been able to identify its presence in known Egyptian texts.

Keeping in mind these three clues of golden—to guard, to protect, and the generative function, I came across another clue from the culture of ancient China. Of all the ancient cultures in the world no two were more isolated from each other, as far as current knowledge tells us, than ancient Egypt and ancient China. In ancient Egypt there is no known documentary evidence of any belief about a sacred mushroom. However, in ancient China there is a definite legend and a belief about a mushroom which is called the plant of immortality. I traced back the sources of this myth.

The use of a special (and unidentified) mushroom was believed by the ancient Taoists to confer all the gifts of immortality upon the one who could find it. The most ancient Chinese character that I could find for the word "mushroom" is to be found in the no-longer-used, archaic, Chinese hieroglyphs. These dropped out of use near the beginning of Chinese history, probably before 2000 B.C. The Chinese hieroglyph for mushroom from this archaic period is:[1]

The first sign represents a tree or wood, and shows the roots of a tree, as well as the branches above the ground. This has the phonetic value LIN. However, this sign has a determinative value and is not pronounced phonetically in the word for mushroom. The presence of a tree sign in the word for mushroom in ancient China corresponds to the hieroglyphic sign given in the word for mushroom by the Ra Ho Tep personality.

The second sign in the Chinese is based on the hieroglyph for the sun. The hieroglyph for the word sun is pronounced JE. However, neither of the phonetic values given for each of these signs gives the word for mushroom. Instead, the Chinese have a unique word for mushroom which is EUL. The word EUL conveys no overtones of a sacred plant, but does picture a tree sign and sun sign to portray the mushroom.

However, in the currently used Chinese characters for "the plant of Immortality" we have the word, LING CHIH.[2] LING in this context means spirit, or divine. Does LING derive from the archaic heiroglyph for tree, LIN? We do not know. CHIH definitely means mushroom. Now we are able to draw a comparison between the Chinese

[1] Won Kenn, *Origine et Évolution de l'Écriture Hiéroglyphique et de l'Écriture Chinoise.* Université de Lyon, Paris, 1939.
[2] C. A. S. Williams, *Outlines of Chinese Symbolism and Art Motives.* Kelly and Walsh Ltd., Shanghai, 1941, p. 116.

142   THE SACRED MUSHROOM

words given for a mushroom, and the Ra Ho Tep personality word for a mushroom.

Both languages make use of a hieroglyph for a "tree" in the representation of a mushroom.

Both languages incorporate a property of the sun (golden) in the hieroglyphic representation of a mushroom.

Both ascribe sacred or spiritual values to the mushroom.

Such a correspondence cannot be accidental and demands further investigation. China has been subject to repeated invasion by peoples from northern Asia. Since Siberia is the heartland of the related beliefs of the "traveling soul" and the mushroom cult, the cultures in this area bear closer examination.

The Koryaks in the northeastern part of Siberia have been reported by Jochelson to use the *Amanita muscaria* in order to have the traveling-soul experience. This practice is reserved for a special clan type of priest called a shaman. In this report by Jochelson I found the answer to my original quest for a connection between shamanism, the out-of-the-body experience, and the *Amanita muscaria*. Another group of people to the west of the Koryaks carry on the same ritual practice. These are the Yenisei Ostyaks. My interest in these people was stimulated by the remarkable word that they have for the *Amanita muscaria*; they call it HANGGO.

Now this word is of great interest for our quest, in that it appears to be related to a whole family of Indo-European words for the mushroom. Following the philologist Pokorny, Wasson has shown that the root word for a cluster of words used for mushroom is GWOMBHO. This is the ancestor word for fungus as the following list by Wasson shows:

> Slavonic: GOMBA
> Gothic: WAMBA
> Hindi: KHUMBI
> German: SCHWAMM
> Russian: GUBA
> Greek: SPONGOS
> Latin: FUNGUS

These words for the mushroom are all within the Indo-European language group. As we go from the West to the East across the northern Eurasian land mass, from the Finno-Ugric language groups to

THE SACRED MUSHROOM   143

the Ural-Altaic language groups, we find the following words for the mushroom:

Mordvines: PANGGO
Cheremissian: PONGGO
Vogul: PANGKH
Ostyak: PONGKH
Yenisei Ostyak: HANGGO

The affinity between the two lists of words from different language groups is suggestive of some unknown common root word. Rather than seek for a common philological root, I chose to try to relate these different words around an ideology centered about a sacred-mushroom practice. I wanted to find if there was a cluster of words which described the main features of shamanistic ritual, the mushroom, and the idea of the traveling soul.

I first noticed that the Mongol word for the spirits of the shaman is ONGUN. This word, completely unrelated to the Indo-European root for mushroom, nevertheless shows phonetic similarity. I point out the Spanish word for mushroom, HONGO, as suggestive. Could the word for mushroom, HANGGO, as used by the Yenisei Ostyak shaman, have any relation to the word for spirits, ONGUN, used by the same people?

In this quest I turned up another clue from the language of the Eskimos. The core of the Eskimo culture is based on the practice of shamanism. I found that the Eskimo word for a shaman is ANGE-KUT. The Eskimos are believed to be a North American group who have only a few hundred members living in Siberia, and these are supposed to have crossed the Alaskan straits in recent times.

I now had a group of words which sounded surprisingly alike, clustered around the *shaman-Amanita muscaria-spirits* axis, with which to begin a game of combination and permutation. I arrived at the following arrangement:

Yenisei Ostyak:	H ANG GO	*Amanita muscaria*
Mongol:	ONG gUN	Shamanistic—"Spirits"
Eskimo:	ANG eKUT	Shaman (controls spirits)
Ra Ho Tep:	ANG KHUT	
	AAK KHUT	Sacred Mushroom
Chinese:	LING CHIH	Sacred Mushroom
Vedic Sanskrit:	AHI CHAT TRA	Mushroom

144   THE SACRED MUSHROOM

The last is the oldest word known in the Sanskrit language for a mushroom. AHI means a snake, or a cow; and CHATTRA means a parasol or umbrella. We have noted before that the hieroglyphic sign in ancient Egyptian which looks like a mushroom is translated as parasol or umbrella by Egyptologists.

This little family of words which I collected is held together only by their place in the chord of the overtones and undertones of the sacred-mushroom ritual. The words come from unrelated language groups and a common root word is unknown. If I could find some common root from one of these language groups that carried all the meanings which are listed above, I might be in a position to decipher the meaning of the hieroglyphic word presented to me by the Ra Ho Tep personality. The other possibility of a solution lay in the assumption that there existed some freemasonry of the mushroom cult which had distributed a set of practices and words across the boundaries of diverse cultures separated widely in space and time. I had to explore both possibilities.

It will be noticed that I have separated each of the above words into two parts. I was led to this device by an inspection of the Greek word for a messenger or spirit, AGGELOS, from which we get the word angel. The stem of this word is, ago, and means to lead or to guide. From this stem is derived words for "messenger" in other Indo-European languages, and its philological root is AG.

My first task was to trace the words that branched out from the root AG. The root AG means: to drive (something along), either by driving it from behind or leading it in the fore. From this root AG come words which mean: to travel (Nordic AKA); a messenger (Latin ACTUS); a wooden staff to drive cattle along (Sanskrit ASTRA); and a place to which one drives cattle, a field (Latin AGER). The basic idea is that of motion under compulsion, as symbolized by the staff of authority. It is curious that this dynamic concept should lead to a static concept like the Latin word, AGER, for a field. I examined this word more closely.

AGER is the root of an old Latin word for a mushroom used by Pliny, AGARICUM. This word is still in common use to designate certain mushrooms as the "agarics," and the *Amanita muscaria* is called the fly agaric in textbooks today. In order to find out if the agaric was in any way related to the root AG as it might apply to the *Amanita muscaria*, I turned to one of the earliest references on the subject, *The Syriac Book of Medicines*, edited by Sir E. A. Wallis

THE SACRED MUSHROOM 145

Budge. This manuscript was probably written in the twelfth century of our era, and contains prescriptions from Greek, Egyptian, Persian, and Indian sources. That the text reflects medical ideas existing long before the Christian era is shown by some of the prescriptions which bear a resemblance to those in the Ebers Papyrus of ancient Egypt (1700 B.C.).

In this Syriac text I found four medical prescriptions, each of which contains as an ingredient the "Agarikon Fungus." The diseases that these prescriptions were supposed to cure could give one a clue as to the pharmacological properties of the drugs, particularly, the Agarikon. Prescription Number 1 (page 47) is supposed to "bringeth on the menstrual flow." This can be compared to a modern drug, Prostigmine, which has some of the same effects as muscarine. Prescription Number 2 (page 48) checks "excessive flow of urine." This is a characteristic effect of the *Amanita muscaria* which I have observed, namely, that while one has the need to urinate, the production of urine has been decreased, and by contraction the bladder is empty. Prescription Number 3 (page 145) restores strength in "paralyzed nerves." This again is consistent with a muscarinic action of the *Amanita muscaria*. Prescription Number 4 (page 275) is a preparation for "immortality."

The first three prescriptions would fit the expected pharmacological actions of *Amanita muscaria* and its drug, the muscarine. The fourth prescription is reminiscent of the Chinese prescription of the "Pill of Immortality," and the "Plant of Immortality." Certainly this Syriac text strengthens the belief that at this historical period, and perhaps earlier, the Agarikon refers to the *Amanita muscaria*.

At this period of history, the Lebanon Mountains of Syria were covered with huge forests of oak, pine, and cedar. We recall the famous "cedars of Lebanon" of the Bible. In the earliest period of Egyptian history the Pharaoh Snefru sent fleets of ships to this country to secure the precious cedar wood for ship and temple construction. All the evidence indicates that in ancient Lebanon (then called Syria) the natural conditions of forest, moisture, and climate were ideal for the growth of *Amanita muscaria*. The Amanus Mountains are to the north of this chain of mountains, but I have never been able to find any link between Amanus and *Amanita*. It is certain, however, that the Syrian physician knew about the Agarikon fungus, and it is highly probable that it was the *Amanita muscaria*. If this is

146   THE SACRED MUSHROOM

so, it is conceivable that his source of this drug would be in the Amanus or Lebanon Mountains of Syria.

Thus our search for a basic meaning for the root AG has led us in a wide circle to the Agarikon in Syria. The HANGGO of the Yenisei Ostyaks is definitely known to be the *Amanita muscaria.* While the philological relation between AGARIKON and HANGGO is not known, we certainly can relate these words in our family of sacred-mushroom ideological terms.

I now summarized the word cluster that I had derived from the Indo-European root AG which applied to a clarification of the hieroglyphic sign for a ladder, AAK, given by the Ra Ho Tep personality in the hieroglyph for mushroom. The ideas inherent in the root AG— to drive, to lead along, to travel, and messenger—are also present in the symbol of the ladder sign AAK as used by the ancient Egyptian scribes in the literature of the Pyramid Texts. This root AG is directly related to the old Latin (AGARIKON) and Arab (AGHRIKON) words for mushroom. The Vedic Sanskrit, AHI, in AHI-CHATTRA (mushroom) is derived from the root AG and means a cow (an animal that is driven along).

In the case of the Chinese, LING, meaning spirit, or spiritual, in the word for "sacred" mushroom, LING CHIH, LING may well be derived from the archaic LIN, meaning wood or tree. Although the lexicons do not state this, it is my opinion that the meaning of the root AG is to be sought in the image of "driving, or leading (people or animals) by means of the wooden staff as a symbol of authority."

The wooden staff of the shepherd, the scepter of the Pharaoh, and the ebony wand of the magician are rather obvious symbols of authority and power. But I believe that the ancients and primitives attached still deeper meanings to such staves. Wood was the source of artificial heat and light by fire. The ancient god of this function in India is Agni. Wood, however, was more than firewood—it was also the residence of the tree spirit. Such tree and wood worship of resident spirits from many parts of the world is amply illustrated in *The Golden Bough.*

Perhaps this immanence of spirit in wood was based on the observation of a natural phenomenon like the glow (of static electricity) from the tip of a sentry's spear on a dark night. Such eerie light observed on the tips of the masts of ships, and from trees, must have excited profound feelings of awe and mystery in the mind of the ancients. It is not difficult to understand why wood and trees were

THE SACRED MUSHROOM 147

deified, and why they were believed to contain spirits. And when the giant bolt of lightning smashed a mighty oak, no man could avoid profound fear and respect for the power of Zeus or of Indra.

We must also remember that the ancients and primitives firmly believed that lightning engendered mushrooms. If they were of a culture that also believed that the sacred mushroom could send the soul flying out of the body, how awesome must have been in their minds the combination of the power of the lightning and the mushroom! The ancients were well aware of the dependence of mushrooms on trees for growth. Since both the tree and the mushroom were the receptacles of mighty spirit power, how natural to link the two together! Perhaps this is the reason why the tree and the mushroom are found to be coupled in the hieroglyphs of ancient China, as well as in the hieroglyphs of the Ra Ho Tep personality.

The immanence of gods in trees is not unrelated to the history of the Hebraic-Christian tradition as brought out in a remarkable insight by Hrozný.[8] This scholar has deciphered some of the inscriptions of the Proto-Indian culture (ca. 2900 B.C.) of the Indus River Valley in India. One of the deities discovered by Hrozný is called YAYASH, YAE, or YAVE, a protective god whose emblem was a tree, and the tree seems to have been his abode. Hrozný states that "the name is derived from the Indo-European root of the verb *ei-, *ia-, Latin—eo, Old-Indian—Yäti, meaning he goes, he drives; the Hittite—ya (yattari) meaning he goes, he marches. The name Yayash may mean walking, going, marching, a pilgrim, which is an appropriate word describing the sun's daily journey westward."[4]

The name of the Yayash deity seems to have originated from hieroglyphic Hittite (the Turko-Syrian area is their homeland) rather than from Aryan. Hrozný goes on to identify Yayash with the Old Testament god Yahveh. He states that Yau is the Semitized form of Yayash. According to Exodus, Chapters 2 and 3, Moses became acquainted with the god Yahveh while staying with the priest Jethro on his flight from Egpyt. Yahveh appeared to him in a burning bush, which Hrozný relates to the sacred tree of the Proto-Indian god Yayash. Hrozný states: "To interpret the god Yahveh of the Old Testament from the Proto-Indian inscriptions from Mohenjodaro and Harappa seems to be a paradoxical idea which I present to the experts

---

[8] *Ancient History of Western Asia, India and Crete*, Bedřich Hrozný. Philosophical Library, New York, 1953, p. 180 ff.

[4] *Ibid.*, p. 180.

148   **THE SACRED MUSHROOM**

with all reserve."[5] The sacred animal of Yayash seems to have been the ox.

I have felt it necessary to include this digression because it points to the existence of a god of the tree in the Syrian territory in the early part of the Third Millennium B.C., whose sacred animal is the ox. We shall find a reference to a similar god in this area from Egyptian sources of about the same period. (See Appendix 3, page 208.) Since the Indo-European languages have illuminated the meaning of the root AG, we now turn to the other syllable of our mushroom complex, namely, KHUT.

[5] *Ibid.*, p. 182.

chapter XIII

Since the Indo-European languages had been fruitful in word products of the root AG for our mushroom quest, I made the Sanskrit word AHI-CHATTRA (mushroom) the beginning of my analysis of the second syllable of the group of mushroom words. In the Sanskrit, CHATTRA means a parasol or umbrella. The Egyptian word for parasol is given by the hieroglyph ⌐ pronounced SHU, or SHU.T, and later, under the XVIIIth Dynasty, is given as KHAIBIT. I have already given reasons why this Egyptian hieroglyph should be considered as a determinative for a mushroom in certain passages. The Indian usage clearly associates this image of a parasol with a mushroom.

There are cognate words in Arabic, Hebrew, Syrian, Ethiopian, and Egyptian for CHATTRA, and they all mean cover, clothe, shelter, umbrella, or hut. This leads to the Indo-European root for these words which is GEU, or GU. The basic meaning is that of a bow, curve, or arch. The picture of covering something readily comes to mind from this prime meaning. This root word includes the idea of an arch like ⌣, as well as the protective or supporting arch ⌢. According to Pokorny[1] the root GU has nine distinct suffixes, and we will give selected examples from each suffix group that form a constellation of mushroomic ideas.

The root GUD in Sanskrit as GUDAM means viscera or intestines, and in old English the same root yields KITE—belly or stomach. In

[1] *Indogermanisches Etymologisches Wörterbuch*, Julius Pokorny. A. Francke AG. Verlag, Bern, 1951.

149

150 THE SACRED MUSHROOM

Swedish KUTA is a humped back, and KOTT is the pitch or resin of the pine. In Anglo-Saxon CYTE is a hut or a house. In Old Nordic and Low German KUNTA and KUTTE respectively denote the female organ, Latin CUNNUS.

The root GUGA has the basic meaning—ball. In Polish GUGA is a boil or pustule, and in Lithuanian GUZAS is a knot, boil, or tumor.

The root GUPA has the basic meaning, hole in the ground. In Greek GUPE is a cave, hawk's nest, or a secret hiding place. In English we get the words COVE and CAVE from this root. In Dutch KUIF is a crest, the top of the head, or the top of a tree. In Old Nordic KUFR is a roundheaded staff or club.

The nasalized variant of GUPA is KHUMB. Pokorny tenders the opinion that the meaning of this root is surmised from KOBEN (pigsty) combined with GUFRA (deep) leading to the idea of burying a thing, secret. The Sanskrit KHUMBHIKA means a pot or a jar. The root word from Sanskrit for mushroom, or scabrous growth, is GUBA from which comes the Hindi, KHUMBI; Sindhi, KHUBHI; Punjabi, KHUMB—all meaning mushroom. In Anglo-Saxon CUMB is a bowl or a shallow valley.

In Lithuanian GUMBAS is a swelling, or a knot. In Lettish GUMBA is a swelling or a boil. This leads to the many words for mushroom already known for northern Europe.

In the last root derived from GU, namely, GEU, we come to the same root given by Hrozný as *EI for the tree god Yayash. This root means he goes, he drives. This is not to be confused with the basic meaning of AG which is to drive something along. In Sanskrit JUNATI means to drive along in a hurry. In Old Nordic KEYRA means to travel.

Thus from this root GU, we can derive a composite meaning for our mushroomic stem (KHUT, GO) which unifies the other aspect of mushroomic lore. We have the basic ideas of a vessel (cup), a shelter (hut, parasol), a bulbous enlargement (club, boil), a mushroom, and the idea of travel (*EU).

Using this skeleton of words clustered around the Indo-European roots AG and GU, I have been able to arrange a cluster cognate in the ancient Egyptian language. For example, the phonetic stem AK, or AG (they are interchangeable at times) in Egyptian is not etymologically related to the Indo-European root AG, but a similar cluster of ideas surrounds it. The symbol of authority, the wooden

THE SACRED MUSHROOM   151

scepter, is called AK in Egyptian. The idea of walking or striding over terrain is contained in a word of the same sound AK, but built up of different hieroglyphs.

The sound AAK in Egyptian as the noun is a ladder, and as the verb means to ascend. AKAT is a servant or messenger, and AKHU is the imperishable spirit of man which goes forth to dwell with the gods. AAK is also a master or a mason, and one of the names of the god Tehuti is AK.R. Although there is no record of the mushroom in ancient Egypt, a bulb-ended plant like the leek is named AAKT. Thus we find a cluster of Egyptian ideas around the phonetic stem AK which resembles the ideas clustered around the Indo-European stem AG.

The Indo-European root GU has yielded a series of words built around the idea of a positive- or negative-curved jar shape. These are words meaning curve, arch, hut, hump, bone joint, club, viscera, female sex organ, cave, jar, and mushroom. In the ancient Egyptian we can find a cognate set of meanings built around the phonetic stem K. For example, a curve or a bow is AK. An arch is KAPT, and the roof of a house is KAP. For a bone joint we have KS, and for a hill KA. The viscera or intestines is given by KAB, and the female sex organ is KAT. KRT is a hole in the ground and KRTA is a cave dweller. A basin or a bowl has many words in Egyptian, one of which is GAA. There is no word similar to the Sanskrit GUBA (pot, jar, mushroom). But there is a word for an alabaster jar, KB, which is interesting if we recall that the Egyptians pictured an alabaster jar as a vessel from which the gods dispensed "life" (ANKH). GU is a sweet-smelling plant, and KU, or GU, means to ascend.

We can now summarize our long excursion into word roots and mushroomic clusters. Both the Indo-European root AG and the Egyptian phonetic stem AK yield cognate clusters of words and ideas. The basic idea is to drive or lead something (under the direction of a staff of authority). This is applied to the person who is led or driven —a servant or a messenger. This leads to the concept of a messenger, spirit, or an angel in both language groups. Here the action is the important element.

Both the Indo-European root GU and the Egyptian phonetic stem K mean a curve or a bow. This gives rise to words for various curved objects such as a jar, hut, cave, female genitalia, club, and hill. In these ideas it is the shape of the object that is the key element.

152   THE SACRED MUSHROOM

Both Indo-European roots AG and GU lead to words for the mushroom. These Indo-European roots combine uniquely in the Greek word AGGELOS, meaning messenger, as the root of the word, angel. The phonetic stems AK and K in Egyptian do not lead to a word for mushroom since this is unknown in this language; there is no certainty that the word AAK KHUT is combined from the two stems to yield a word for mushroom. It is my opinion that the two Egyptian stems under consideration combine uniquely in the word HKA which means prince, ruler, or magic. Although HKA is commonly translated as magic, we do not know what the Egyptians really meant by this word. I believe it will be found that HKA can be clarified in its meaning if approached from the beliefs and ideas associated with the sacred-mushroom practice.

We now make a trial grouping of our key words arranged so as to combine their meanings and phonetic elements in such a way as to bring out the sacred-mushroom terminology.

Greek:		AG	GOS	A Vessel
Egyptian:		ANK	H	A Vessel (of Life)
Greek:		AG	ON	Council, Congregation
Mongol:		ONG	UN	Shamanistic Spirits
Egyptian:		AK	KO	Council of the Wise
Eskimo:		ANG e	KUT	Shaman
Yenisei-Ostyak:	HANG		GO	(*Amanita muscaria*)
Egyptian:		AAK	KHUT	Ladder (Mushroom?)
Chinese:		LING	CHIH	Mushroom of Immortality
Greek:		AG	GELOS	Messenger, Angel
Egyptian:		AK	HU	Imperishable Spirit
Egyptian:		H	KA	Magic Power

The left-hand column lists a group of languages distributed over the entire Northern Hemisphere. The middle column lists a cluster of sounds produced by the human vocal cords that show great similarity but have no unity on the basis of philological considerations. The right-hand column, a key list of terminology from the shaman-mushroom-spirit cult, gives the sounds in the middle column a unique cohesion. I am sure that a similar cluster of sounds could be artificed for other ideas, but I know of no such wide-ranging cluster that is so homogeneous for the complex practice represented.

With this guide map of words and sounds before me I began to examine the large body of ancient Egyptian hieroglyphic texts to see if I could find any hint pointing to the existence of a sacred-mushroom practice. The most likely sources of information would be the many funerary texts and the medical papyri.

In my search I found that the more ancient a text was, the more likely it was to contain hints and clues that furthered the quest. Of the medical papyri, *The Smith Surgical Papyrus* seemed to be the most promising. It was written in the seventeenth century B.C., and according to the opinion of Breasted this medical document was probably already in circulation at the time that the Great Pyramid was being built (ca. 2700 B.C.). This was helpful in that it coincided with the date of the historical Ra Ho Tep personality, and with the date of the earliest Pyramid Texts.

On June 16, 1954, the Ra Ho Tep personality had described the preparation of a mushroom unguent that was to be applied to the top of the head at the site of the anterior fontanel, or the "soft spot" (on the head of an infant). Curiously enough the Buddhists of India and Tibet call this same spot the Aperture of Brahma and believe that the soul exits from the body at this point. We have no record of a similar belief in ancient Egypt. But in *The Smith Surgical Papyrus* we find a word that describes the anterior fontanel as AḤT, or weak-place. This word is not found in any other Egyptian document. Another rare word in this text is UḤNN, which is descriptive of the crown of the head. If the Ra Ho Tep personality was actually reporting a mushroom practice in ancient Egypt then I should be able to find some mushroomic allusions to the crown of the head, or the weak-place.

In the Egyptian book entitled *The Ceremony of the Opening of the Mouth*, a funerary text giving instruction for the proper spiritualization of the deceased body, we find a vignette for the Ninth Ceremony which shows:

154 THE SACRED MUSHROOM

This is a mushroom-shaped sign placed over the weak-place, AḤT, of the crown of the head. No explanation is given for this representation in the accompanying text. Egyptologists have interpreted this as the shadow of the deceased following the umbrella or shade interpretation of the

sign.

The Smith Surgical Papyrus (page 524) uses another word for the crown of the head, which is of more common usage, UPT. UPT is represented by the picture of a set of ox horns, and the Tep sign, . A word derived from UPT has the set of ox horns combined with the door sign spelling AP, and this means to open or free. It struck me as curious that the top of the skull, a pair of ox horns, and a door sign should be combined to form the verb, to open, or to free. This made sense only by considering the closed top of the skull as containing an important opening such as is found in the concept of the Aperture of Brahma. This suspicion was confirmed by finding that a few more signs added to the root signs—the ox horns and the door—yield a word UPU-TAU which means messengers. A variation of this word by the addition of the SHU sign spells UP SHU which means to brighten or to illuminate.

Following this clue furnished by the ox horns on the top of the head, and the door sign (as the root AP), leading to UPU-TAU— messengers—and UP SHU—to brighten or illuminate—I found a rare word from the Pyramid Texts, PR, which means to come out. But the word PR as used in line 1107a in the Pyramid Texts means to ascend (to heaven). Thus this series of signs and words which refer to the crown of the head clearly suggests that the Egyptians had some such concept as the Aperture of Brahma as an exit point for the soul from the body.

My word-guide list turned up another promising lead in the quest for the presence of a mushroom cult in Egypt. The idea of fields (flat fields) in the Egyptian word AHT naturally brings to mind the meeting of such fields with the sky, or the horizon. The Egyptian word for horizon is AḤT and this comes from a root AH which means to be beautiful or glorious (like the sun rising in the East). In the Pyramid Texts we find that AḤU is sunshine, and AH, or AKHU is a blessed spirit. The pyramid of Khufu is called AAKHUT, or "The Pyramid of the Horizon." These ideas of horizon, sunshine, glorious, and spirit are combined in one word in line 4b of the Pyra-

THE SACRED MUSHROOM 155

mid Texts in the name of the god HR·AH·TE, Harachte, or Horus of the Horizon. He is a form of the sun god when rising or setting, and was associated with Ra. Fundamentally, Harachte, in the form of a falcon, was the ancient sky god Horus manifested in Ra. This god is associated with the East, and it has been suggested that his origin lies in Syria. Here we have another pointer toward Syria.

In a picture from the Book of the Dead, page 253 (Budge), Harachte is shown as a falcon standing on a mushroom-shaped standard. This form of standard is found elsewhere in the Egyptian texts, and its interesting feature is that it has two stalks, one shorter than the other. This could not help but remind me of Wasson's description of the importance which the Mexican *curandero* attached to pairing off his ceremonial mushrooms by some intuitive sense so that each member of the pair was matched against the other. The *curandero* then eats these pairs of mushrooms as though they were balanced against each other in some way. This may reflect some intuitive knowledge of a balance of drugs in each of the mushrooms, just as there is a balance of atropine and muscarine in the *Amanita muscaria.*

Many of these clues which I found all pointed toward the East, and particularly to Syria. The K root stem which I had found in the Egyptian language for the second half of the mushroom word turned up as FNHU, or FNKU, the Egyptian word for Syrians. From this word the Greeks derived the word Phoinikes for Syrians, which then became Phœnicians. The chief port of Syria was called KPNY, or KBN (during the Old Kingdom) by the Egyptians. The Syrians themselves called this city and port GBL, which the Greeks corrupted to BYBLOS. Byblos was famous for the papyrus manufactured from its papyrus swamps as writing material in ancient times, and this form of paper came to be called "byblos" by the Greeks, from which we get our word for the Bible, which means book. From this word for the Syrian city of Byblos (GBL) the Egyptians derived a word KBNW which means "something" connected with magic. This derivation excited my curiosity—what was this something that had to do with magic?

I found that the Semitic-Arabic word for mushroom is GBA. Could this word for a mushroom have any connection with the Syrian city GBL, and the word for something magic that the Egyptians connected with the same city which they called KBN? I have already described the Syriac medical use of a mushroom highly suggestive of the *Amanita muscaria,* and the fact that the Syrian coastal range area, that is, the Amanus and the Lebanon Mountains, offered ideal growing conditions for the *Amanita muscaria*. The Egyptians from the earliest times knew the forests of the Lebanon in Syria as their workers cut its timber for ship and temple. If a mushroom cult existed here, where all signs point to an abundance of mushroomic life, could the earliest Egyptians have learned of it and then taken it back to Egypt? In order to answer this question I had to seek for some evidence of such a cult in this region of ancient Syria.

Montet discovered the ruins of an ancient Egyptian temple in Byblos. Frankfort has emphasized the high antiquity of this temple and dates it back to pre-dynastic times, or before 3000 B.C. He justly raises the question as to why the Egyptians should have maintained a temple so far from the Nile, especially in a territory that did not belong to them. He has given no answer to this puzzling problem. Frankfort, also in this article,[2] shows a seal carved on the rare green jasper to which he assigns a Syrian origin, and compares to seals found at the site of this old Egyptian temple in Byblos. He dates this seal to the first intermediate period, or the period following the Old Kingdom period of Egypt. I reproduce this seal here for its interest.[3]

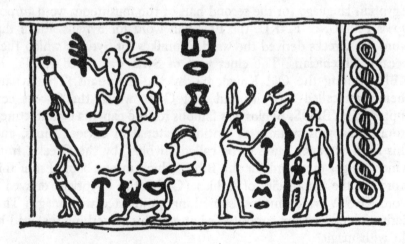

The scene is essentially Syrian for this early period, but shows heavy Egyptian influence, as evidenced by the use of Egyptian

[2] *Journal of Egyptian Archeology*, H. Frankfort. Vol. XII, Parts 1 and 2. April 1926, pp. 80–99.
[3] *Ibid.*, p. 92, Fig. 7.

THE SACRED MUSHROOM   157

hieroglyphs. The scene on the right commands our attention. The central figure is a priestly character with the headdress of the god Horus, or HRAHTE. In his hand he offers an object that looks like the double ax, or a mushroom, to the supplicant on the far right. Above the two human figures is a hare. The supplicant before the priest-god holds in his hand a staff which is of the same form as the DJAM, or UAS, scepter of Egypt. Between the two figures are some hieroglyphs. The ones directly under the double ax, or mushroom, are RA N RA, and I believe that this can be translated as: in the name of Ra. Closer to the supplicant are two hieroglyphs. The top one is unknown in the Egyptian language, but has been compared to the ANKH sign. I do not believe that this is the ANKH sign but that it is an approximation of the middle hieroglyph in the Ra Ho Tep personality writing for AAKHUT. The sign below this unknown sign is the reed which stands for the letter A, and may be read in conjunction with the sign above it as AAKHUT. The entire phrase reading: in the name of Ra, [the] AAKHUT.

This reading is given because it is believed that the object offered by the Horus priest is the mushroom, as the AAKHUT. This is confirmed by other considerations. The two eyes of Horus are the sun and the moon. We have already shown the presence of the sun, as Ra, in this seal. The hare above the mushroom is the symbol of the moon in the Aryan cultures, and the Syrians were heavily influenced by their Aryan Hittite neighbors to the north. I believe that the hare symbolizes the moon in this seal as the counterpart of the sun, the two being the left and right eyes of Horus.

By referring to the line 445a of Appendix 3, page 193, the reader will note that a passage from the Pyramid Texts expressly states that an unknown object, an AAKHUT, is made or fashioned by the god Knum. The couchant lion with the head of a man, the flat horns of a ram, and the two plumes above the horns in the Syrian seal is a representation of the god Knum. Knum as a god of the waters is also represented by a jug. While the basin here shown above the lion is not the type of vessel usually shown with Knum, I believe that the Syrian craftsman was not too careful in a precise representation, but has shown a more simple form of vessel. The couchant figure of what I believe to be Knum is gazing directly at another mushroom representation, and this is in harmony with the Pyramid Text passage (line 445a). The DJAM, or UAS, staff held in the hand of the supplicant is also the staff that the god Knum is often represented as

158   THE SACRED MUSHROOM

holding. Even the mushrooms are present as a pair. This constellation of factors leads me to believe that the vignette we are viewing is that of a sacred mushroom rite from ancient Syria.

The other animals in the scene, the three Horus falcons, the couchant ibex, the monkey-imp, and the bird (as the freed soul?) are well-known figures in Syrian representations. The three hieroglyphs at the top middle of the scene are difficult to interpret since they are not drawn exactly like known Egyptian hieroglyphs. They could designate the name of Horus, or they could mean the temple of (unknown sign).

The entire scene which I have described and interpreted when taken into the Syrian context previously portrayed (medical, climatology, flora, and the words AGARIKON or AGHRIKON and GBA —mushroom) leads me to believe that a sacred-mushroom cult existed in this part of the world in ancient times. However, I have not by any means exhausted the evidence for this conclusion for Syria. Since the northern Syrian area is one of the geographical starting points for the migration of the Indo-European peoples (Hittites, Aryans, Vedic Indians, etc.) I have been able to trace the diffusion of a mushroomic culture into India and Greece in particular. In Syria there have been found stone monuments called Cromlechs.[4] These are made of two stone pillars supporting a mushroom-cap-shaped slab of stone. The monuments look like huge mushrooms on the landscape. They have been found in a huge arc extending from Britain as far south as the Malabar Coast of India.[5] In Europe most of them were destroyed by zealous Christians centuries ago, who looked upon them as undesirable relics of a pagan worship. Their shape and function has never been satisfactorily explained. But this is not a part of our search here.

The discovery of a relationship between the god Knum, and a picture of what I believe to be a pair of mushrooms, led me to examine more closely a passage in the Pyramid Texts which states that an AAKHUT was made or fashioned by the god Knum. The English, French, and German translations of this passage did not mention any mushroom rite. In the face of these standard readings by eminent Egyptologists I approached the translation of this passage in

[4] *History of Syria*, Philip K. Hitti. Macmillan, New York, 1951, pp. 27–29, p. 43.
[5] *The Celtic Druids*, Godfrey Higgins. London, 1829. Pl. 26, 39, 40.

THE SACRED MUSHROOM   159

the ancient hieroglyphs with the utmost caution. I knew that my amateur knowledge certainly could not compare with that of the experts, and that a clear recovery of a mushroom rite from this passage presented enormous odds against success. But I had to face this difficult task in order to reach a conclusion as to the authenticity and validity of the Ra Ho Tep phenomenon.

For the results of this translation the reader is referred to Appendix 3. This passage has never been translated before as a mushroom ritual text. Should my interpretation come to be accepted in time, I believe that scholars will find other long-hidden passages which will reveal other details of a sacred mushroom ritual. For example, in the offering lists of Nefert, the wife of the historical Ra Ho Tep, there is (Plate 15) a rare sign showing a man grinding something with a mortar and pestle. The caption for this scene spells AKHUT as follows:

In this case AKHUT refers to the horizon. It is my interpretation that this may refer to the grinding of the AAKHUT of the Ra Ho Tep personality, i.e., the mushroom.

This interpretation is supported by the hieroglyphs from an entirely unrelated passage in the Pyramid Texts, Utterance 244, line 249. This rubric is usually entitled "Breaking of the Two Red Jars" (Mercer). It is true that a pair of red jars is depicted, called G·G, but their determinative is that of a man grinding some particles (of an unknown product) with a mortar and pestle. When we recall the pot or jar root of the mushroom words, their presence in pairs, and other mushroomic allusions (grinding the warts, the color red, or golden, etc.) it seems that it would be well for Egyptologists to reexamine such passages.

In the pyramid complex of King Zoser, who reigned about a century before Snefru, were found six stelae illustrating certain phases of the HEB-SED ceremony. It is believed that this ceremony was concerned with fertility beliefs and rejuvenation rites on behalf of the King. The stelae are of interest because they show pairs of umbrella-shaped figures being held up behind the King. Furthermore, the ancient Egyptians in these six portrayals clearly distinguished between the form of a sunshade, a lotus, and what I believe to be the mushroom. The following figure shows the mushroom forms illustrated in

one of the stela.[6] That these are mushrooms can be inferred from the fact that Plate 17 clearly shows a single sunshade being held up by an ANKH sign,[7] and Plate 41 clearly shows two lotus leaves, each being held up by an UAS scepter.[8] Below the two mushrooms of Plate 16, one supported by an ANKH and the other by an UAS scepter, are two red crowns of the kind drawn by the Ra Ho Tep personality. The UAS scepter is seen to be of the same form as the scepter portrayed earlier in the Syrian scene. Under each crown is the SHEN, symbol of kingly authority. This sign is also the symbol of another name of the god Knum in the form SHENU. Thus the two red crowns of SHENU symbolize the power of "two red crown plants of life" as previously found in the Ra Ho Tep personality writings. And since two mushrooms are clearly portrayed we can conclude that these are the red crown plants of life. (See Appendix 4.)

In the HEB-SED temple court of Zoser were found quantities of fine red-polished potsherds of the Old Kingdom together with flint knives and bones. Such a find is of interest in connection with the rubric mentioned above.

Although there will be some controversy, and much time will pass, I am reasonably sure that diligent scholarship will eventually uncover further confirmation of the hints given by the Ra Ho Tep personality for the existence of a sacred-mushroom cult in ancient Egypt.

[6] *The Step Pyramid*, Cecil M. Firth and J. E. Quibell. Cairo, 1935. Vol. II, Pl. 16.
[7] *Ibid.*, Pl. 17.
[8] *Ibid.*, Pl. 41.

chapter XIV

I HAD done everything within my means and ability to rule out the possibility of fraud or delusion on the part of Harry Stone during the manifestation of the Ra Ho Tep personality. I was finally satisfied, after three years of investigation, that he was an innocent agent in the transmission of intelligence in the ancient Egyptian language. The negative business of analyzing Harry was a simple job compared to the task I now had to face. What is the meaning of this case history?

This case was fully discussed with many people over four years, and some of the proposed explanations can now be considered. The explanation given most often was that Harry had some deep dark motive and had fabricated the Ra Ho Tep character and writings. The other was that he had unconsciously learned this Egyptian material, and then unconsciously remembered it in trance. I feel that I have adequately disposed of these explanations. A question was raised occasionally as to whether I had fabricated the entire case. But there are too many witnesses to this manifestation who know that my role was strictly that of a critical observer.

To those who believe in the reality of extrasensory perception, the Ra Ho Tep manifestation was a case of simple telepathic transfer. The hypothesis was advanced that Harry was in telepathic communication with an Egyptologist somewhere in the world and that this was the source of his knowledge of the Egyptian language of 2700 B.C. This explanation, as a matter of fact, has not yet been ruled out, and it may be that the publication of this work will bring to

161

162   THE SACRED MUSHROOM

light some scholar who is preparing a thesis on the existence of a sacred-mushroom ritual in ancient Egypt.

Other "telepathic" hypotheses stated that Harry got his knowledge of the Egyptian language from the person that was present. But there were a number of people, each of whom was present alone with Harry, when the Ra Ho Tep Egyptian language was produced. These were Betty Stone, Alice Bouverie, Henry Jackson, Wim Jochems, and I. None of us knew the ancient Egyptian language when first faced with this case. I was the only one who eventually studied the language, and it was three years before I felt that I had mastered even the rudiments. In addition to these people there were at least two dozen other individuals who at one time or another witnessed the production of the Egyptian writing in the trance state by Harry Stone. None of these people was familiar with this language.

While the Ra Ho Tep personality was being manifested I tried to induce three different Egyptologists to witness the phenomenon, but they were too busy to get involved. This was regrettable, as it weakened our analysis of the case in regard to the telepathy hypothesis.

Another explanation offered was the reincarnation theory. It was stated that Harry had lived a previous lifetime as Ra Ho Tep, and that he was now recalling its memories. I did my best to investigate this possibility, and this was one of the reasons behind the hypnosis study. It must be emphasized that the Ra Ho Tep personality insisted several times that both Alice Bouverie and Betty Stone had lived once before as his contemporaries, as Antinea and Nakita, respectively. It would have been logical to attempt deep hypnosis on both of them in order to check this possibility. This was not done because both Alice and Betty resisted the idea of being experimental subjects. And if a person is antagonistic to the idea of being hypnotized, it is rather hopeless to begin such an experimental venture. Neither Alice nor Betty was intrigued by the idea of having lived a previous lifetime, and this lack of interest kept them in a rather objective frame of mind about the Ra Ho Tep case. When Alice did slip into a hypnotic state by accident on July 4, 1955, she gave directions as to where a mushroom could be found. This incident did not reveal any Antinea personality or any Egyptian material. Its remarkable feature was that it helped us to find a mushroom of the description given independently by the Ra Ho Tep personality a year earlier.

The hypnosis studies with Harry ruled out any significant personal

THE SACRED MUSHROOM   163

experience on his part with advanced Egyptian learning in his lifetime. His contact with Egyptian material in museums was casual and did not account for the type of material produced. The fragments of the Egyptian language that he did produce under hypnosis have a number of explanations. By the reincarnation theory he could have recovered this from his own memory storehouse of a previous lifetime, but this was not related to the Ra Ho Tep personality. Or he could have recovered this memory from the previous Ra Ho Tep trance manifestations which he had experienced. Weakening this argument is the fact that he produced one phrase that he had never given before. Or he could have slipped into a deep trance and manifested the same Ra Ho Tep personality as on previous occasions. The hypnosis studies did not establish a sound case for the reincarnation theory. This theory was weakened by Alice's independent and involuntary production of intelligence which led to finding the golden mushroom advertised a year earlier by the Ra Ho Tep personality.

This synchronism of independent events extended into other areas. On August 15, 1953, Wasson obtained firsthand evidence of the existence of a sacred-mushroom cult in Mexico. This knowledge, as far as I have been able to determine, was not shared by Harry, Alice, or myself at this time. It first came to my attention February 12, 1955, and became public knowledge with the *Life* magazine article of May 13, 1957. The Ra Ho Tep personality first manifested on June 16, 1954, and announced in a veiled way the existence of a mushroom practice in ancient Egypt. The mushroom described by the Ra Ho Tep personality was not of the species found in Mexico, nor of the red *Amanita muscaria* known in Europe, or known to be used ritually in eastern Siberia. In New York City it was a yellow, or golden, mushroom that was described in English by the Ra Ho Tep personality, and a red mushroom in hieroglyphs. The golden mushroom was found a year later in Maine and subsequently in Massachusetts. I myself at the time had no special interest in mushrooms of this type, but independently of the Ra Ho Tep information had become interested in this problem by my experience in the Army. I would say that the main reason that I became interested in a study of Harry Stone was the quest for a drug that would amplify extrasensory perception. These independent events of the years 1953 to 1955, extending from Mexico to Maine, present a remarkable synchronism.

Were these relations all accidental and artificially formed into a pattern after the facts were in? If we knew the true relation of all

164   THE SACRED MUSHROOM

these events we might have the answer to the meaning of the Ra Ho Tep manifestation.

Arising out of this synchronism of events we have two objective groups of data to consider. The first is the hints about a sacred-mushroom ritual in ancient Egypt and the evidence of the ancient Egyptian language produced in written, vocalized, and sign-language form.

Up to the time of the Ra Ho Tep information, neither Harry, Alice, nor I, as far as I can determine, had any knowledge of the existence of a sacred-mushroom cult. That this was known to a small group of scholars in the world before this time is a fact. That Harry was not in possession of such information I am quite certain. But of all the knowledge which I now know to be extant on this subject, I have yet to find a single reference that suggests that such a practice was known in ancient Egypt. As far as I know, the Ra Ho Tep personality was the first to announce such a possibility. In checking this information in the ancient Egyptian literature, I have become convinced that there is evidence from the texts themselves that a sacred-mushroom ritual was practiced in secrecy in ancient Egypt. I believe that this evidence exists in the archaic HEB-SED ceremony and in the later Pyramid Texts. It is for Egyptologists to check the validity of my suggestion. This is a problem of historical research that can be settled in time.

The second group of data concerns the effects of the sacred mushrooms. From the Mexican source we are led to believe that besides the hallucinogenic effect of the mushroom there is conferred on the user some divinatory ability. Mr. Wasson kindly supplied me with a small number of the Mexican mushrooms which he collected, and I was able to do a few experiments with them. My results were negative, both as to hallucinogenic effects and as to influencing extra-sensory perception scoring rates.[1] Because of the absence of hallucinogenic effects, it is my belief that the mushrooms, which were several months old, had lost their potency, and that therefore my experiments have no validity. Further study is indicated in this direction with the Mexican species of sacred mushrooms.

---

[1] It must be emphasized that the Mexican species of mushrooms are not poisonous, and that our specimens of *Amanita muscaria* are poisonous. Hence one could take large numbers of the Mexican mushrooms (I took twenty-two at one time) in order to bring on the hallucinations. One could not take the same liberties with the *Amanita muscaria*.

THE SACRED MUSHROOM   165

In testing my own specimens of *Amanita muscaria* for their effect on extrasensory perception, I found that they had no enhancing effect on normal individuals, i.e., individuals who were not able to achieve significant scores on laboratory tests for telepathy. The only two individuals that I had access to who were proven sensitives, Harry Stone and Peter Hurkos, did respond dramatically to *Amanita muscaria* administration. I have already described these effects and have pointed out the difficulty in interpreting the results. It is certain that their sensitivity was not depressed by the mushroom and that both exhibited euphoria—inebriation in one case, and an ecstasy or trancelike state in the other. These effects followed the administration of very small doses of the mushroom drug. The effects of massive doses were not studied because of the danger of poisoning and because the hallucinogenic effects are well known from the literature. My conclusion from these observations is that this drug if properly used in sensitives offers promise for further research in extrasensory perception.

The Ra Ho Tep personality had given me the impression, both from the verbal and written statements, that the mushroom had been used in his "time" to fission the soul safely from the body during life. This idea can be interpreted in a number of ways. This could mean that the mushroom was used symbolically, as is the golden bough, in the passage I have quoted from *The Aeneid*,[2] or as a symbol of initiation into the greater mysteries in which some "soul experience" is dramatized. In this interpretation the mushroom would not be an active agent in this experience.

The mushroom may have been used, either symbolically or as an unguent, in a ceremony similar to that found in the Tibetan Book of the Dead, where the purpose is to achieve the proper separation of body and soul at the moment of death. This would be a funerary use of the mushroom, and the passage that I have translated from the Pyramid Texts may be interpreted as such. But my impression of the Ra Ho Tep personality statement was that the mushroom was to be used as an operational fission agent during life.

In this case the mushroom would be used as a part of the HEB-SED ceremony. This ceremony existed in Egypt from prehistorical times. It is believed to be based on the primitive practice of the sacrifice of the King. Since kingship was sustained in virtue of magic

[2] See Appendix 3, p. 201.

166 THE SACRED MUSHROOM

powers, it has been suggested that the King may have escaped sacrifice by proving anew his magical powers and physical prowess by undergoing the ordeals of the HEB-SED ceremony. It is known that a part of this ceremony involved prolonged physical exertion by running and by dancing. If the King survived this trial he would have proved his physical prowess. In keeping with the thought pattern of primitives and the ancient Egyptians it is also conceivable that he had to get the consent of the spirits or the gods in order to maintain his office. It is here that a magical mushroom would fit into the HEB-SED ceremony. The King would use it as an active agent to fission soul from body and assay the perilous journey to the underworld and secure the permission of the gods for continued kingship.

The King is known in this ceremony to have been accompanied by a priest of the "Souls of Nekhen." The souls of Nekhen were the preceding and prehistoric kings of Egypt. It may well be that the Egyptians did not trust the report from the underworld made by the King, and sent the priest along to confirm the visit to the gods and to report independently their verdict. Such an interpretation would satisfactorily account for the participation of the priest in the ceremony with the King. It would also account for the presence of the sacred mushroom in the HEB-SED ceremony. A test of this hypothesis would lie in the existence of a similar practice elsewhere, and in checking its operational validity by experiment.

I looked into this proposition very carefully as a possible avenue of research. I would like first, however, to define terminology in order to present more clearly my ideas in this direction. I see the problem primarily as one in which it is desired to have the body in one place on the one hand, and the principle of consciousness in another place on the other hand. This idea is amply illustrated by the personal dream experience which I have described. I would term this general phenomenon a mobile center of consciousness, or MCC for short.

In this concept the body of a person, conscious or unconscious, would be in a definite place. However, the conscious awareness of that person would travel to any other point it desired, see what is going on, return to its body, and then report on where it has been and what it has seen. This intelligence could then be checked against what actually happened at the time or place allegedly visited. The MCC would have to travel to the place of interest with a sense of being all of a piece, lacking only the physical body. I will henceforth

THE SACRED MUSHROOM   167

use this definition, and the term MCC, in my discussions of what I have been loosely calling the "out of the body" experience, "traveling soul," etc.

The only place in the world where I have been able to find a tradition of artificially inducing an MCC by means of a mushroom is in eastern Siberia, among the Koryaks, Chukchi, and the Tungus peoples. I would like to summarize this practice in a composite form from the individual practices of each of these groups.

The candidate for shamanship, either male or female, usually receives a "call" to this form of priesthood somewhere between the ages of fifteen to thirty-five. He then retreats into the woods from his clan or tribe and lives a solitary and ascetic life for a long period. During this time he indulges in fasting, exposure to cold, and physical exertion to an extreme degree, bordering on masochism. He has two objects, one to master the physical man and the other to master the "spirits." When he feels that he has achieved mastery over both he returns to his clan and undergoes testing at the hands of the elders to see if he is master of himself and the spirits.

The physical tests are prodigious. For example, among the Tungus, nine holes are cut in the ice of a frozen river and the candidate has to jump into the first hole and surface in succession at each of the remaining eight holes. If he survives this test he goes on to other extreme tests, finally being allowed to demonstrate his ability to call and control the clan spirits. Having passed all these tests and shown his ability in divination, medicine, soothsaying, and spirit mastery, he then becomes a full-fledged shaman.

One of the great feats of shamanship, and one which is not attempted too often, is one I describe as an MCC. This involves several days of preparation before the ceremony and includes fasting, meditation, and exposure to hardship. The real ceremony begins at sundown in the clan tent with the clan assembled. A coniferous wood fire burns in the middle of the tent, saturating the air with smoke, carbon dioxide, and the terpene and camphor compounds of combustion. The shaman begins the ceremony by beating on a special drum (and this may go on for several hours), improvising chants and hymns to the spirits that are to be called to assist in the separation of the soul from the body. At a certain point in the drum beating he gets up and begins to dance vigorously. Others take up the drum beating. His dancing may last for hours. When he begins to get exhausted he stops to consume large amounts of certain intoxicants.

168 THE SACRED MUSHROOM

These can be alcohol, tobacco, or the *Amanita muscaria*, depending on the culture and the clan. Inebriation, in any case, sets in.

Now the shaman resumes his dancing and physical exertion with renewed effort. Travelers have reported seeing the shaman leap as high as six feet in the air over and over again while in this state. The purpose of the shaman, apparently, is to achieve complete and utter physical exhaustion. When he has reached this stage he suddenly collapses on a specially prepared couch and passes into a deep trance. The tent becomes silent, and the shaman is carefully watched to make sure that he does not unexpectedly die. Shamans have been reported to die from the great exertion.

In this exhausted trance state the shaman alleges to have the experience of the mobile center of consciousness, and travels to the spirits of the underworld or to distant points on the earth. When he awakens, he is gently revived by tea and by deer meat, and he then reports to the expectant clan what he has seen and what he has experienced.

This general procedure is followed by most shamans, with minor variations from tribe to tribe. Even the Tibetan oracle follows many of the same procedures, but we are told that it is forbidden to use drugs in Tibet.

Assuming that this is the method which the tradition of thousands of years has found to be the best for inducing at will the state of an MCC, I realized that this was a form of experimentation that was truly heroic and one which I could not embark upon lightly. I therefore looked for some less drastic technique of achieving the same end. It has been reported that success had been achieved in attaining an MCC by the use of hypnosis, and I turned to this method first.

The experiments were carried out with three different hypnotists and several selected subjects. None of the subjects achieved a mobile center of consciousness. Even the addition of the mushroom to hypnosis did not achieve an MCC. I soon realized that trial and error by the hypnosis technique was not the answer. In order to do this experiment properly one would have to screen thousands of individuals in order to find a few ideal subjects for the try at an MCC by hypnosis. This would have to be done by some large organization with adequate funds and many subjects at its disposal.

I have come to the opinion that the only satisfactory method of attempting to achieve an MCC is to follow the arduous traditional practice of the Siberian shaman, using the fasting, motivation, ex-

THE SACRED MUSHROOM 169

haustion, and mushroom techniques. The first requisite will be to carefully select a few subjects. In addition to being young, physically hardened, and of good intelligence, they will have to exhibit sensitivity of the order shown by Stone and Hurkos. With such subjects and the experimental facilities of a modern research hospital, I believe that the attempt could be made to achieve an MCC without endangering the life of the aspirant. I fully believe that such experimentation will be attempted within the next decade, and I hope to have a share in bringing it about. Until such an experiment is organized and given sufficient time for a fair trial, we will not have an answer to the question as to whether an MCC is a reality or a long standing delusion of the human mind. Obviously, the use of the "sacred mushrooms" in such an experiment would be only a small component of the total technique.

I do not want the reader to get the impression that I think that an MCC can be achieved solely by the use of drugs or physiological and psychological techniques. My experience leads me to believe that these by themselves are not sufficient. What these other techniques are needs some elaboration.

All of us feel that we know what prayer is as a personal experience. Nowadays there are many Christian church groups who believe, and practice, that prayer is a consciously directed intelligent operation that can increase plant growth, heal the sick, and integrate the mind and the soul. Whatever the truth may be about prayer I have the feeling that it does so work, and have seen striking examples of effects proceeding from it. It is my opinion that whatever reality lies behind operant prayer, and I cannot define it, it is this reality that should also be a part of the technique which I have proposed for the MCC experiment.

A consideration of the kind of reality known through prayer brings us back to the question of what is the reality behind the Ra Ho Tep manifestation? With this reality I could not experiment—I could only eliminate factors ascribable to fraud, error in observation, or delusion. After all this detective work was done, there was still an unaccountable residue of reality. For this I listened to many ready explanations from my well-meaning friends. In addition to the ones I have cited, some felt that it was a simple case of mediumship on Harry's part, with a spirit controlling his actions. Reincarnation-minded friends, who cannot accept the hypothesis that spirits can be disincarnate, feel that Harry is an example of a very old and deep personal memory manifestation. Friends who cannot accept either hypothesis, but who accept extrasensory perception, feel that this is an unusual case of clairvoyance with a sort of super-fixation on early ancient Egypt. Another offers the hypothesis that this is a case of spirit possession. In this case "spirit" means to him some sort of mental debris without personal form that acts as a poison on the mind. Another friend offers the "collective unconscious" hypothesis, stating that the language phenomenon is a sort of common race property which has surfaced in Harry's mind under stress, like a well-formed but unusual mandala.

It is difficult for me to accept most of these suggestions directly from the stock shelf of psychological and religious ideas. I fully accept the idea that there was present behind the façade of the Ra Ho Tep personality intelligent direction. My question is: What is the nature of this intelligence? Is it a finite soul related to Harry or to the historical Ra Ho Tep? If so, is such a soul capable of making a mushroom grow on a five-hundred-square-yard portion of the earth's surface after predicting its coming two days before the event? I feel that our diligence at the moment showed this to be the only such mushroom to be found within several square miles of this area. It does not seem likely that it was a fortuitous coincidence.

What created the apparition that Peter Hurkos saw twenty-four hours before Alice's death? What made Hurkos give a psychometric reading, on a piece of paper written by Harry Stone's hand, that described, not Harry Stone, but a scene from ancient Egypt? Why was the sacred and secret nature of the mushroom cult revealed in Mexico in 1953 and not in 1553? Why was a similar cult in ancient Egypt (this is yet to be accepted) revealed in New York City in A.D. 1954 and not somewhere else in 1954 B.C.? I feel that one must look for a pattern behind these events, an intelligent pattern, rather than a study in isolation of a finite Ra Ho Tep personality.

It is true that the Ra Ho Tep personality did appear to be finite and consistent as evidenced by statements made about a personal life, consistent use of an archaic form of the Egyptian language, and some faint identification with a historical Ra Ho Tep. As such I am willing to accept tentatively his existence as a disincarnate personality who may have lived once in ancient Egypt.

But it is difficult for me to understand why, if he died forty-seven hundred years ago, there is no other record of his manifestation before or elsewhere. And why was he limited in expression by Harry's

THE SACRED MUSHROOM 171

knowledge of the English language and his personal knowledge of ancient Egyptian? It almost appeared that he had been in some deep-freeze state in the years intervening between his death and current manifestation. In this sense he was a fossil of a ghostly world.

My true feeling about the meaning of this case is that there is a reality to intelligence in the cosmos that is independent of its manifestation in finite bodies or finite souls. In other words, intelligence is substance, and ideas may be a table of intangible elements. Just as we ascribe evolution and entropy to biological and physical processes, so there may be a concomitant evolution and extropy of ideas that appear in the human race. These ideas appear in the human mind in the form of images, symbols, archetypes, or in personalized form. They break into the human sphere at multiple points, as, for example, the independent invention of the calculus by Newton and by Leibnitz. When such synchronism of ideas occurs in many different humans, we call it *Zeitgeist*, or a spirit of the age. In interpreting this phenomenon some take the point of view that the new development is organically inevitable, and sooner or later someone will make the discovery; and others take the point of view that a completely new idea has come to the human mind by an outside influence. In other words, some think of a physical evolution and others think in terms of a psychical teleology. This issue does not become acute until one is faced with a problem like the Ra Ho Tep case.

This brings up an ancient problem which has not yet been resolved; the reception of omens and messages from the beyond. Such great thinkers as Socrates, Dante, Descartes, Schopenhauer, etc., have had their daemons, familiars, or messages. It seems to me that we beg the greater question when we assign such intelligence from beyond the human sphere to another set of finite intelligences, such as daemons and spirits. In all my experience with mediums and such, I have yet to hear any intelligence from such sources that strikes me as being different in quality from that coming from human beings that I have known. In other words, what one hears from alleged "spirits" has all the limitations that go with the finite human mind or intelligence. One is but the echo of the other, and I do not know where the origin lies.

I would prefer to consider both carnate and disincarnate intelligence as being but poor reflections, to borrow Plato's analogy, of a more universal field of intelligence. And of this, we must admit, we know very little. For that matter we know very little of the basic processes and ideas of human minds. The human mind daily goes about accumulating knowledge and classifying it in millions of small ways, and yet we know nothing about the agent of this work. How we grasp and utilize any idea we do not know. Yet we go on using ideas with the greatest sense of importance all of our lives. The basis of creativity in ideas, imagination, reason, and memory is all unknown. It appears to me that if we could solve the process of one function of the human mind, such as memory alone, we could then begin to understand the rest of the idea process and in time the very nature of intelligence itself.

I do not doubt that disincarnate intelligences exist, any more than I doubt that finite carnate intelligences exist. To me they are but opposite sides of the same coin. If we ever get a complete understanding of one, I believe that we shall also understand the other. But in order to understand one of them fully, we have to face squarely the existence of both phases and draw our understanding by earnestly studying both, and thereby move toward the underlying reality.

From my point of view the heart of the problem is to be reached in experiments whose purpose it is to challenge boldly the concept of a mobile center of consciousness. If the MCC does not exist as a reality, independent of the human body, then an age-old source of confusion will have been removed. If science does affirm the reality of an MCC, and I believe the evidence indicates that it will, then we can begin to comprehend intelligently matters which up to now have been largely in the hands of faith. All the necessary techniques are now in the hands of medical science to safely ask this question of nature. All that is lacking is some intelligent enthusiasm. It may well be that the real meaning of a Ra Ho Tep manifestation, and others like it, is to kindle such enthusiasm.

appendices

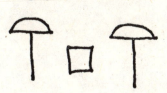

APPENDIX 1

RESEARCH at the Round Table Foundation for the past six years has shown that under special environmental conditions many sensitives demonstrate an increase in accuracy and in the amount of intelligence collected by means of telepathy over that found under normal room conditions. The special environmental conditions are created by placing the sensitive inside of a cubical metal enclosure, or room, called a Faraday Cage.

The Faraday Cage is a copper enclosure whose walls prevent electromagnetic waves and electrostatic effects from passing this metal barrier to the inside. If a radio is taken inside the Faraday Cage it would continue to play as long as the door of the cage is open. But as soon as the door of the Faraday Cage is closed, and providing there is no electric wire passing through the walls, the radio will be completely cut off from the broadcasting station. The radio waves are stopped by the copper wall, and cannot penetrate to the interior of the Faraday Cage.

Likewise, if an electric charge is placed on the walls of the Faraday Cage a person inside will feel no electricity—even if he touches the inside walls of the cage.

The increase in the telepathic prowess of a sensitive was measured by rigidly statistical standards. The method used to measure increase or decrease of telepathic ability was a matching problem test. A familiar example of a matching problem is tossing a coin in the air and calling out whether it will land heads or tails. In other words, in order to be correct one's call must always match the coin face that turns up. This principle is used in almost all tests for ESP. We utilized such

176 THE SACRED MUSHROOM

a test which is called the Matching Abacus Test, MAT for short. It consists of two matching sets of ten different pictures. Each set of pictures is placed in a row. Both rows of pictures are shuffled, and the trial matching of ten pictures is called a run. The pictures are placed under an opaque screen so that the sensitive (the receiver in telepathy) can handle them but cannot see them. The receiver is also blindfolded, so that he cannot see either the pictures or the sender. The pictures are placed in clear plastic boxes so that the receiver cannot touch the surface of the picture, but the sender can clearly see each one.

The receiver places his left hand on one picture in the row closest to him. The sender now knows the picture that the receiver will seek in the other row. The receiver then passes his right hand over the other row of pictures and attempts to find the mate to the picture under his left hand. The sender, by telepathy alone, tries to influence the receiver to pick up the correct picture. When the receiver makes his choice he picks up the plastic box and places it opposite the one under his left hand. Each such matching decision is called a trial. If the two pictures correctly match it is called a hit. A complete experiment consists of five runs, or 50 trial matches. The chance score for a telepathy MAT experiment is six hits out of fifty trials (6/50). The score significant for telepathic ESP is 11/50 (P=.01, or odds that such a score could result by chance once in a hundred such experiments).

A number of different telepathic teams averaged 12/50 hits under normal room conditions (P=.006, or odds that such a score could occur by chance six times out of every thousand experiments). Such a score is considered significant evidence for telepathy.

Inside the Faraday Cage

The same telepathic teams were then placed inside of the specially constructed Faraday Cages. These had no electrical charge on them, and the cage in which the two subjects sat was grounded to earth. The average score of the telepathic teams jumped to 25/50 correct matches. This represents a statistical probability of 1.29341×10^{-10}, or odds that such a score could occur by chance approximately once in ten billion experiments. This is a highly significant increase in ESP test scores, and indicates an increase in the telepathic interaction.

There was a question that the ESP test scores might have increased because of the psychological stimulation afforded by working in a novel environment. In order to rule out this psychological factor the subjects were placed in the cage as before, and it was found that if the wire connection between the cage and earth was broken by a switch the ESP test scores dropped to the level found for normal room conditions. Many tests of a similar nature proved that it was not a psychological factor that was responsible for the increase noted in telepathic ESP test scores.

It is important to record that, under the conditions cited above (associated with scores significant for telepathy), while the receiver correctly matched two pictures, he rarely knew what the picture was. In other words, if he correctly matched two pictures of a ship, he did not know consciously what the picture was. Such manual matching, without knowing which picture was being matched, means that no intelligence was gained from this kind of telepathic process. The correct acts were carried out at an unconscious level of mind.

When the subjects performed inside of the same Faraday Cage but now carrying an electrical charge of twenty thousand volts direct current negative on the outer walls, the scores jumped to an average of 43/50 hits ($P=10^{-26}$). This is a highly significant increase in ESP test scores over those obtained in a Faraday Cage grounded to earth and carrying no electrical charge on its outer walls. But more important than this increase in ESP test scores was the fact that the two subjects were suddenly able to transmit and receive symbolic intelligence.

This means that not only could they match the pictures correctly, but they were able accurately to name or describe the picture. Furthermore, they were able for the first time to read each other's thoughts. The subjects retained this latter ability when placed in separate Faraday Cages 0.3 miles apart. The content and acuity of intelligence gained by psychometry from photos or handwriting was also increased by working inside the charged Faraday Cage.

It must be pointed out that when both subjects are in one cage separated four feet from each other across a table there is a good probability of sensory clue exchange. But the control tests with an ungrounded, or floating Faraday Cage (where the scores averaged 12/50 hits) showed that such sensory leakage was not serious. It was obvious that if the subjects were placed in separate rooms, or in separate buildings, there could be no sensory leakage. But such a separation would also eliminate the precise synchronization of the sender's telepathic "signal" with the receiver's hand when it was over the correct matching picture.

Under the conditions of such separation each subject had only one row of the MAT pictures. The receiver was to rearrange the pictures in his row in the order in which he thought the sender had arranged his row (or the order in which the investigator had arranged them for the sender). Under these test conditions, and when each subject was working under normal room conditions, the scores averaged 6/50, or pure chance-expectation. However, as long as one subject (acting as receiver) was in an electrically charged Faraday Cage he averaged 14/50 hits ($P=4.71$ x 10^{-4} or odds that such a result could occur by chance four times in 10,000 experiments). This score is significant for telepathy. The drop in scoring level is attributed primarily to the loss of precise synchronization between the sending and receiving of telepathic signals. When both subjects were in separate charged Faraday Cages at a distance of 0.3 miles from each other the scores rose to an average of 36/50 hits—showing an increase in telepathy.

Such stabilization of telepathic communication under rigid test conditions made it possible to demonstrate publicly extrasensory perception. An investigating committee of the Psychic Research Society, Massachusetts Institute of Technology, witnessed a telepathic experiment at the Round Table Laboratory. With elaborate precautions against fraud or error the telepathic team of Hurkos and Stone achieved an ESP test score of 18/50 hits ($P=4.01$ x 10^{-6}, or odds that such a score could occur by chance four times in a million experiments).

This represents solid evidence for the reality of telepathic communication between these two men. It must be emphasized again that in our experience, and under such stringent test standards, significant scores are possible on demand only when one or both of the sensitives is inside of an electrically charged Faraday Cage.

APPENDIX 2

Drawing No. 1. *June 16, 1954*

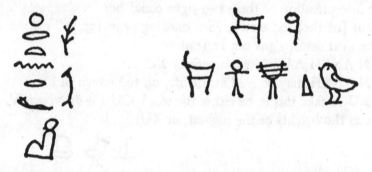

On the left side is a poorly drawn Ra Ho Tep with REN to be possibly read under it as:

(name of). The sign under this:

is not too clear and may have been intended as the F:

REN·F means my name. This group of signs would then mean: Ra Ho Tep is my name. The seated figure under this is too poorly drawn to be given any definite value.

180   THE SACRED MUSHROOM

The SUT and the UAS (or DJAM) scepter appear to the right.
The signs:

are difficult to interpret. Since the subject, Harry, had
been speaking in English a few moments before of a plant, and a
mushroom in particular (see Figure on page 16), it may be that the
sign:

refers to a plant in general.

The red crown coupled with this sign (interpreted as a plant)
makes sense only if one takes into consideration the fact that the
species of mushroom portrayed on page 16 is either red or gold in
color,[1] depending on where it is found. If this be accepted, then a
loose interpretation of these two signs could be:

plant [of the] red crown. (See drawing page 158.)

The next set of signs can be read as:

N ANKH AAK (determinative) KU.

N ANKH may be read as: of life, or, red crown of life.

KU—I take this to be either the word K3U (*Wörterbuch* V, 4)[2]
given as the heights or the highest, or KU:

(*Wörterbuch* V, 8), whose old meaning is not known, and later became a kind of offering. In the Pyramid Texts line 335 and line 949
this word is used as: to ascend [to heaven in a boat]. In keeping with
this latter usage we have present the ladder determinative and therefore the same idea of ascension is portrayed. My translation would
be: ascension. Since red symbolizes victory over life I would translate
the entire group as:

red crown plant [of] ascension [over] life.

---

[1] Species: *Amanita muscaria*.
[2] *Wörterbuch der Aegyptischen Sprache*, Erman-Grapow, Academie Verlag,
Berlin, 1955.

THE SACRED MUSHROOM 181

DRAWING No. 2. *September 4, 1954*
   The Ra Ho Tep Nesu formula (not shown) is quite clear.[8] The writing:

   The first two signs are as in the preceding figure, but the ANKH is more clearly drawn.
   The next four figures are not in linear order, and I feel that they should be rearranged as:

   I have rearranged them on the assumption that we have the same word as the KU of the preceding drawing. With this interpretation I assume that the poorly drawn figure (the second):

is probably:

   This interpretation and moving the sign into this new position give us the word K3U, ascend. The next two signs I would group as:

and interpret the second sign as:

value NM (*Wörterbuch* II, 267). This group would then be the word NMT, to move oneself, or, of persons showing signs of life as opposed to that of death.

[8] Also called the RḤ-NSW·T title or formula. (See page 42 for the drawing of this formula.)

182 THE SACRED MUSHROOM

This group:

$$\text{𓂀 𓊖 𓏲}$$

is, I believe, to be read in the reverse order as:

$$\text{𓏲 𓊖 𓂀}$$

I interpret the second sign as:

$$\text{𓊖}$$

MR, (Gardiner, 518, U23). Thus the word is SMR, friend, courtier, (*Wörterbuch* II, 102) or beloved, most often with a title. Thus the entire line would read:

EN ANKH K3U NMT SMR

or, red crown of life ascension returns friend

or, more freely, a friend returns [with] the red crown of ascension [over] life.

DRAWING No. 3. *August 29, 1954*

The entire drawing is covered with many figures, some of which make no sense. I was not present when they were drawn by Harry, so I do not know the order in which they were written, nor the grouping. From this drawing I select the following group:

The Ra Ho Tep NESU REN formula (not shown again) I would interpret as:

In the name of a friend of the King, or Ra Ho Tep, my name.

The sign in the cartouche may be KNU, but I do not know what this name means. See drawing 4 for an interpretation.

It may also be read as AAKU, the wavy line being the silent determinative interpreted as "waters of heaven" (see footnote 3, p. 133). In this reading, AAKU is the name of the mushroom personalized as a god.

This writing is extremely curt and cryptic, and I feel must be interpreted in the context of the two previous drawings.

The ladder sign has already been established in association with the idea of ascension in the word KU. In this context it appears to be a determinative for the signs:

△ ⌷

I believe that this represents the word GŠ, written also as KŠ (Wörterbuch V, 156) meaning a plant which grows in watery places. But the Wörterbuch does not have these signs containing a ladder as a determinative. Hence we must look upon this grouping as a unique word. Combining the meaning established for the ladder, ascension, and the signs for a plant, KŠ, we have an original reference to a plant [of ascension] that grows in watery places. This coincides with the description given by Harry for his sacred mushroom in other contexts. Perhaps if the original textual source of this particular word could be found we would have the answer to the natural origin of this hieroglyphic writing.

The red crown sign does not seem to have any phonetic value and should be interpreted literally as a red crown. This is plausible in view of the known color of the *Amanita muscaria* as being either red or golden.

The last sign I interpret as UR, great, rather than as a quail chick. The preceding sign is probably TA, offering. The entire line would read:

EN KŠ (ladder determinative) TA UR

or red crown plant offering great

or, more freely, great offering [of the] red crown plant.

To this may be added the meaning of the beetle sign as coming into being.

An alternative translation is possible for this group:

PHONETIC VALUE: N   AAK   KH.T   (Det.)   UR
EGYPTIAN WORDS: EN   AAKKHUT   UR (determinative offering)
TRANSLATION: OF AAKKHUT   GREAT OFFERING.
(or) The great offering of AAKKHUT.

I have chosen the above values because of the probable presence of the word AAKKHUT, which the Ra Ho Tep personality associates with a sacred-mushroom ritual. While AAKKHUT literally means the wood of the ladder, it has been used both in the vocalized and written form in other contexts as referring to the sacred mushroom.

Another possible translation would use the pyramid sign as the determinative for a pyramid. In this case the translation would be: from the Great Pyramid AAKKHUT. Since this is the name of the Great Pyramid of Khufu at Giza, known as The Pyramid of the Horizon, it would give us a dating for the IVth Dynasty. If one makes the assumption that Harry learned the name of this pyramid in his lifetime, then the question arises as to why he used this unorthodox spelling for it. The equivalent spelling used makes it highly unlikely that this writing came from Harry's unconscious memory.

DRAWING No. 4. *August 29, 1954*

I cannot clearly make out the signs which begin by spelling the formula, Ra Ho Tep NESU. The grouping of the two RA signs, the double reeds, and the duck, SA, may be indicative of the formula RA SA, or son of Ra, duplicated. Or perhaps it may mean Son of Ra, twofold? The other six hieroglyphs on the left are more clearly drawn:

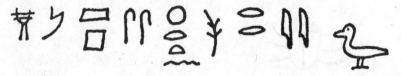

I believe that two words are present here, and the second one appears to be ŠPSS, meaning exalted, venerated, etc.

The first two signs offer difficulties, and their values are respectively K and M. If they are read as MK, this could be a contraction for another word given in Wörterbuch II, 33 for a ladder as:

or M3KT. If this is plausible, the line would read: exalted ladder.

However, since the purpose of these writings seems to be concerned with a plant, and the plant is identified with a ladder, we may use these clues to pursue another meaning. The word K3M is written in a number of ways in Wörterbuch IV, 106, and one of these is:

KM, meaning, in general, a growing place, a garden. Since KM means garden, and our unknown word has this phonetic value, it may be that the plant previously referred to is now associated with some sort of a growing place. The word K3NU has the same meaning as KM (Wörterbuch V, 107), a garden, and more particularly, a vineyard. Since drawing 3 contains the personalized word KNU, there may be a repetition in KM of the idea of a place where plants grow. I would prefer to use the meaning of garden for the word KM, in view of the above considerations. With all reserve, the following translation is offered:

KM ŠPSS

or garden exalted

or, more freely, exalted growing place, a friend of the King, double son of Ra.

Drawing No. 5. August 25, 1954

This set of hieroglyphs is rather loosely drawn, and they offer a problem in determining the intended configuration.

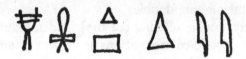

Our serious problem is the second hieroglyph. Its configuration may be clarified by interpreting the other signs first. The word KŠ appears again and may be interpreted as plant, and the determinative may be either the ladder sign as before or the questionable sign between the ladder and KŠ.

The last three hieroglyphs probably mean double offering, or offering of two. It is of interest to note that the only place in the world where there is a well-defined sacred-mushroom cult, Central America, the sacred mushrooms are always used ceremonially in pairs. It is permissible to associate this knowledge with the offering of a pair of two such plants, if they indeed represent mushrooms. In the IIIrd Dynasty tomb of King Zoser the mushrooms are also displayed in pairs.

This leaves our troublesome second sign to consider. It may represent the sign:

(Gardiner, 529, W19), a milk jug carried in a net, whose phonetic value is given as M. If our unknown sign can be equated with this one, then we would have a variant spelling of KM, as found in drawing 4, or garden. In this case the translation would be: double offering [of the] plant [of the] garden.

However, this reading appears to me to be most uncertain. It is also possible that our questionable sign was intended to be:

(*Wörterbuch* IV, 13), S3UA, meaning something that is of two thirds gold, gold, or golden. In this case we would have a repetition of the form used in drawing 3, where the plant is qualified by the adjective red, and in this case by golden. Thus the ladder would be a determinative for the word KŠ, plant, and this questionable sign would be an adjective for it, or golden plant. As stated before, the *Amanita muscaria* is a mushroom that comes in two colors, red and golden. In this interpretation the translation would be:

S3UA KŠ (ladder determinative) TA AIY

or golden plant offering double

or, more freely, double offering [of the] golden plant.

Which of the proffered translations is correct, I am not certain.

DRAWING No. 6. *August 29, 1954*

The left-hand side of the drawing contains the formula written now as: Ra Ho Tep NESU HEN, or the adored friend of the King, or friend of the adored King.

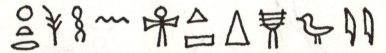

The first sign is not drawn like the previous ANKH and may be compared to the sign:

(*Wörterbuch* I, 204), ANKH, which is described as the name of a vessel in the ANKH form, out of which the gods dispense, or pour, life.

The next group of signs appears to be KŠ, or plant, again. Interposed is the TA sign, offering, as before.

The last three signs offer some difficulty because of the vague appearance of the bird. I believe that this bird can be compared to the one of drawing 4 as a pintail duck with value SA (Gardiner, 471, G39) meaning son. With this choice the group would refer to two sons.

The entire phrase would then be:

RA HO TEP NESU HEN, ANKH KŠ (ladder determinative) TA SIA

or the adored friend of the King offers the life-giving plant [to] two sons.

DRAWING No. 7. *October 5, 1954*

The original drawings are sloppy in places, and I redraw them as I believe the signs and the context warrant:

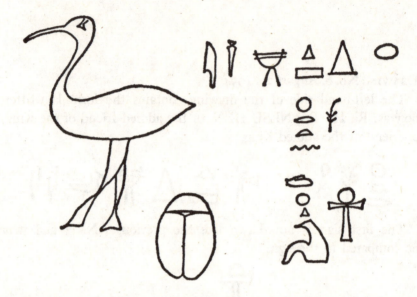

I have tried to clarify certain blurred signs. The sign next to the reed at the top, I believe, was intended to be a dagger rather than a double reed sign. Gardiner states (page 511, T8) that the phonetic value is probably derived from an obsolete word, TEP, dagger, found only once. As TPY it means chief or first. The two signs:

may phonetically be TPY, or they may be AT or ATEP, (*Wörterbuch* I, 141), meaning a loving father, or of a god, as the father of other gods, or Kings. Since the context refers to Tehuti, I would choose the meaning father.

At the bottom right hand, the ANKH had placed on it the △ sign. I believe that this was intended to be placed in conjunction with the signs:

spelling TEKH. Budge (*The Gods of the Egyptians*, Volume I, page 516) gives this as one of the names of Tehuti. This interpretation is confirmed by other signs. The bird on the left appears to be an ibis, symbol of Tehuti. The awkwardly drawn figure below TEKH I believe to be a seated form of the humanized ibis. These three signs all point to the identification of Tehuti in this drawing.

It is to be observed that the pyramidal sign is drawn differently than the previous pyramidal sign for TA, offering. The egg-shaped sign with this new pyramid sign is usually found with a pyramid or temple, as MR. For this reason I believe that this passage refers, not to an offering, but to a pyramid.

The beetle sign is interposed between the ibis and the TEKH, and as such may indicate that it is Tehuti that is coming into being or some such concept of creation.

The word KŠ with the ladder determinative is again present, and would mean plant (of ascension) as before. However, there is a possibility of a variant reading here. If the ladder sign is given one of its phonetic values as AAK, the △ sign one of its possible values as T, and the ▭ sign one of its archaic values as KH (*The Smith Surgical Papyrus*), then this group would give the word, AAKHUT, meaning horizon. Since the line is about a pyramid, this should be interpreted as the Pyramid of Aakhut.

The entire group of signs can now be read as:
RA HO TEP NESU, TEKH ANKH. KHEPER TEHUTI ATEP AAKHUT MER
or, a friend of the King, Tekh [dispenser of] life. Tehuti [coming into being] father [as creator] of the Aakhut Pyramid
or a variant reading, retaining the word KŠ:
a friend of the King, TEKH [dispenser of] life. Tehuti creator [of the] plant [of the] pyramid.

DRAWING No. 8. *September 4, 1954*

This writing was done by Harry (in a trance state as usual) on a tablet of plasticine (artificial clay). It was written with a stylus which he prepared on the spot from a modeling tool of wood (which had been handed to him).

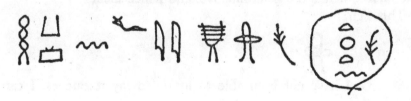

The Ra Ho Tep formula is personalized here and is the only place in which it is used as a name.

I believe that there are only two words present in this line. The first word is made up of the (first) seven signs. *Wörterbuch* III, 175 gives for HK3, when appearing in the genitive or with a suffix, the meaning, magic, or supernatural power of someone or something. In our line it is present with the genitive adjective N and the suffix FY. Gardiner (Par. 76) states that FY is used with nouns having the singular form but having some implication of duality. Thus this word HK3·N·FY implies some duality in the second word.

I cannot find the second word in the lexicons, and one can assume either that it is meaningless or is an original word with some meaning:

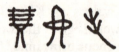

The first sign, the ladder, has been repeatedly used as a sign of ascension and as a determinative for a plant that gives ascension. Harry has described the plant in question as a mushroom of a golden, or a red, color. He has stated that this "sacred mushroom" is supposed to induce a fission of body and mind, so that the mind can operate (or travel) independently of the body. Presumably the ladder is a symbol of such free travel of the mind.

The third sign can be compared to the archaic sign found in the Pyramid Texts for wood, KHUT, or it may be a plant. I have noticed in other contexts that Harry has referred to plants interchangeably as either wood or as plants. If we are dealing with a plant in this context, his use of this hieroglyph would be consistent with his other statements.

The first and third signs taken together are, phonetically, AAKHUT. Since Harry has vocalized this sound a number of times when speaking of his sacred mushroom, I believe that this is the sound and spelling that he refers to. For this reason, I believe that the middle sign is the determinative, and hence silent.

This sign:

I have not been able to locate in my researches. I can

THE SACRED MUSHROOM 191

only compare it to signs that appear similar. In drawing 5 we found a sign:

which I compared to:

S3UA, golden. It is possible to apply this same meaning of golden to the present context, in which case the unknown AAKHUT, either as horizon, or as a mushroom, would be a golden AAKHUT.

However, there are other textual considerations that may amplify the interpretation of this sign. The first word appears to refer to a "magic" quality of AAKHUT. There is a sign which may be compared to our unknown sign:

(Gardiner, 588), whose value is SA, and means magical protection. This would fit our context, except for the fact that SA does not have a ring around it, but has the band shown above.

There is another sign which must be considered which comes closer to our:

and this is:

(Pyramid Texts 452a) and which is used as the determinative for the god, ḤKAS. This god, and some apparel associated with him, appears in connection with the rare divine name, HEPUI, and this latter is given in hieroglyphs by the Wörterbuch III, 71 as:

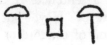

While these signs on each side of the square are usually interpreted as sunshades, I believe that they should be re-examined in the light of what is now known about the use of sacred mushrooms in pairs. The sign □, P, could well represent the door (rather than a mat) by which the ancient Egyptians fancied one entered RESTAU, or the door to eternity. If we interpret the two sunshades as a pair of mushrooms, we find an unsuspected relation between our unknown sign, the determinative for ḤKAS, and the god HEPUI.

While the meaning of our unknown sign cannot be precisely determined, each of the several possible meanings I have cited apply in some measure to the context of a sacred mushroom appellation.

HK3·N·FY AAKHUT

or the supernatural power of AAKHUT

or, more freely, the supernatural power of [a pair] of AAKHUT (i.e., if this latter word is interpreted as a sacred mushroom).

DRAWING No. 9. *December 9, 1954*

On December 9, 1954, there was written by the entranced Harry Stone the above statement in the presence of Alice Bouverie and myself. It was one of those unexpected seizures that lasted only about three minutes while we were sitting and discussing the possibilities of further experimentation with him. It is perhaps the most enigmatic of all of his writings, and I will analyze it sign by sign.

The first sign is a double hash mark. It means you two, and is a sign that replaced human figures on the assumption that it was magically dangerous to give their names.

The next group sign is the double reed, which also looks like knives or feathers and has the phonetic value of AIY. It is very rare to use them as a prefix or at the beginning of a sentence. It can be used as an interjection of address in the sense of "Hey!" but it is more likely that it is used here as an interjection of concord or agreement.

THE SACRED MUSHROOM   193

As the latter it can have the prospective or future reference meaning in connection with the verb. The next sign I take to be the hind quarter of an animal, and this has the phonetic value of PEH. It has the meaning of to attain to, in the sense of a circumnavigation, or to the end of (a cycle). Thus this initial group of hieroglyphs can be translated as: You two, to attain to —— or to the end of ——.

The next three signs form a group, namely the wavy line, and the two hanks of rope. This group of signs means eternity. (*Wörterbuch* II, 299.)

The next two signs may be taken as a group, namely the circle with the radiating lines, and the TEP sign below it. This group could mean light, or the ancient gods, and, as the latter, may refer to the Ennead or the nine great gods of Egypt who were clearly referred to on October 5, 1954, page 128. It can also mean divine cycles or to hear the most holy speak, but in the absence of a NETER sign these renderings are uncertain.

The last two signs in this line are, respectively, the ANKH, or vessel (out of which the gods pour Life), and the swallow UR, which means great.

Thus the first line can be translated as: You two, to attain the eternity of the great (nine?) life——

The first sign in the second row shows three horizontal bars. It is used as a substitute for names which are regarded as magically dangerous to utter or record. It can also be used to indicate the plural. Since the next figure, that of a bird's head, is a determinative, meaning to offer, it is assumed that someone, or some group, is making an offer. Since the identity is unknown, we use the designation X for the unknown one or group who "offers" something. Thus: X offers.

The remaining five signs offer considerable difficulty, since they appear as ideograms and not as words.

The third and fourth figures in this bottom line represent knives. There are many phonetic values for a knife. The Erman Egyptian Dictionary lists thirty-four different names for a knife. From this list I have chosen the value MAS as best fitting the context. One of the reasons for this is that the sun bark of Ra is sometimes represented by two knives. This same sun bark has a certain portion of it represented by the same determinative as our second hieroglyph, namely, the bird head. Furthermore, the sun bark has as its most conspicuous portion two unusual masts which are of the same form as the fifth sign, or the one to the right of the two knives. This very sun

194   THE SACRED MUSHROOM

bark is often depicted in duplicate, as are our knives, and so I feel that our knives are really symbols, for reasons unknown, of the dual sun barks which make the journey of Ra across the sky each day. Thus the third, fourth, and fifth signs taken together may have the phonetic value MANDT, and designate the sacred bark of the sun. Pictorially it is also the determinative for sail when divine journeys are meant.

The sixth sign is that of a rudder and is found only in the oldest religious texts (*Wörterbuch* II, 46). The next sign, that of an inverted branch, I have not been able to identify, and must guess that in this context it may serve to represent a primitive form of rudder. Thus these last two signs may be interpreted as rudders used as determinatives indicating the guidance of this divine journey in the sun bark.

The entire hieroglyphic statement can be tentatively translated: You two, to attain the eternity of the great gods (nine?) of life, X offers guidance on the divine journey.

A variant translation may be made by reading the hieroglyphs from the right to the left:

The divine journey is offered you [two] by the great gods of life to the end of eternity.

Because of the textual difficulties this translation is presented with the utmost reserve.

APPENDIX 3

THE two passages in the Pyramid Texts that caught my attention as being suggestive of a sacred-mushroom rite are known as Utterances 300, and 301. I carefully examined this passage in the original hieroglyphs, and in French, German, and English translations. Only the original picture writing of the hieroglyphs suggests in a veiled fashion a mushroom practice. This is comparable to the problem of the Syrian seal in the thirteenth chapter, page 154, which I have deciphered as a mushroom ritual representation.

These passages describe the ferryman who conducts the deceased across the waters, and may be compared to the classical passage across the Stygian Lake and the ascension of the King to heaven. In interpreting these passages we must remember that the literal translation from the Egyptian to the English is a relatively simple process, but it does not tell us what the Egyptians really meant by these words. The passages are ritual and sacred, and therefore reveal only the surface beliefs of the time. They are also magico-religious, and as such had deeper meanings which the literal words do not convey. We will attempt to bring to the surface some of these deeper meanings as we translate each line.

UTTERANCE 300

LINE 445a.   *To say: O Ḫrti of Nsa·t.*

These lines were recited by a priest at the funeral ceremony, hence the introduction, to say. The god being addressed, Ḫrti, is not well

195

196   THE SACRED MUSHROOM

known. Ḥrti is a god of a city called Nsa·t. The location of this city is not known, but it is presumed to have been near Memphis. Ḥrti is related in other passages to Osiris, god of the dead. He was a clan god in the form of a ram. He is called a ferryman in this passage (see next line), and as such is to ferry the King to the other world.

"*ferryman of the AAKHUT fashioned by Knum.*"

Every word in this line needs amplification. The word which has been translated as ferryman can also mean a fluid of some sort, as well as a swampy place. AAKHUT has been translated by others as the AAKHUT boat, on the assumption that the reading, ferryman, necessarily implies a boat as the vehicle. The god Knum is a ram-headed god, but is well known as compared to our knowledge of Ḥrti. Knum was a god of the water in the earliest times, and later is represented as the god who fashioned men, as the potter fashions clay. Hence his other symbol, the water jug.

The key to the meaning of this line (and to the passage) is to be found in the word AAKHUT. I have already pointed out that this is the word first given by the Ra Ho Tep personality to describe the sacred mushroom, or some rite connected with it. This word appears only in this line in the entire body of the Pyramid Texts in this precise spelling. For these two reasons this passage attracted my attention. Its usual translation as a kind of boat is admitted to be uncertain.[1]

We find this same word (with a modified spelling) in a passage from the tomb of the historical Ra Ho Tep which has never been satisfactorily translated or understood. In his tomb Ra Ho Tep is portrayed gazing off into the distance, and the line containing the word AKHUT is supposed to explain his action. The only clear part of this line indicates that Ra Ho Tep watches the house of eternity or the temple of eternity. The presence of the word AKHUT in association with the house of eternity caught my eye, because the latter means either the tomb or the other world. I assumed that the AK-HUT of the tomb meant the same thing as the AAKHUT of the Ra Ho Tep personality, and based my attempt at translation on this. The original Egyptian is: MA ŠDŠR MU AKHUT PR NḌT. My translation is: Ra Ho Tep watches [guards] the red water vessel [of the] sacred mushroom of the house of eternity.

---

[1] *The Pyramid Texts, In Translation and Commentary*, Samuel A. B. Mercer. Longmans, New York, 1952. Vol. II, p. 207, l. 445a–b.

THE SACRED MUSHROOM   197

Here the AKHUT is described as being the red water, or (vessel of) the red juice of a plant. The AKHUT is also associated with the house of eternity, or the other world. The reader will remember that the Ra Ho Tep personality has described the yellow mushroom powder (from its skin and warts) as yielding a red fluid when mixed with water. I have checked this statement, and found that the powdered skin and warts from both the European (Germany) and the American (Maine, Massachusetts, California) species of *Amanita muscaria* when mixed with water produce a reddish fluid. The hieroglyph for red in the Ra Ho Tep tomb is a flamingo, and its color is well known. In Egypt the color red was also symbolic of victory over life. In many cultures different birds are symbolic of the free soul or spirit. We can now better understand how the ancient Egyptians wrapped their ideas in symbols by the association of the mushroom, red, and the flamingo as indicative of a process whereby the soul is freed of the impediment of a body. The Ra Ho Tep personality emphasizes the belief that the soul can get out of the body by means of the mushroom. And the Ra Ho Tep tomb portrayal amplifies this idea by associating the AKHUT with the house of eternity. (See Chapter XIII, page 157 for the AKHUT from the tomb of Ra Ho Tep's wife Nefert.)

This passage from the historical Ra Ho Tep's tomb is internally consistent with the statements of the Ra Ho Tep personality and all that is known about the sacred mushroom practice. Since the sacred mushroom is a vehicle for the passage to the other world, it is understandable why the word AAKHUT has been interpreted as a vessel in the form of the boat. Such a mistaken notion was no doubt fortified by Knum being a water god, and by further assuming that he fashioned the AAKHUT vessel as a boat. But it is much more consistent to conceive of Knum (as the water god) fashioning (or growing) a mushroom, rather than a boat.

In the light of this interpretation the entire line (445a) becomes internally consistent with a mushroom ideology. This line literally means: water (rather than ferryman) of the mushroom, fashioned by the god Knum. Since Ḥrti and Knum are both ram-headed gods, they may have shared some related function in the growth or preparation of the AAKHUT in the belief of the ancient Egyptians. (See Footnote 2, page 133 for relation between mushroom-shaped plants and water.)

198   THE SACRED MUSHROOM

LINE 445b.   *Bring This to N. N. is SEKER of RESTAU.*

N. stands for the name of the deceased King, and will be used throughout this exegesis of the text. Bring *this* is stated in an emphatic way without naming what it *is* that is to be brought, but it is apparent that *this* is the AAKHUT, or the sacred mushroom. The King is called SEKER of RESTAU. Seker is the oldest god of the dead in Egypt, and he is also identified in this role with Osiris. Seker's boat, which is used to ferry the deceased to heaven, is the ḤENU boat. The Ra Ho Tep personality has drawn a picture of such a boat and has personified its name in a cartouche. However, at the time no explanation of its role was given. RESTAU is the gate or door to the eternity of the other world and at this early time was believed to be located in the necropolis of Giza. Thus this line means that the deceased King is at the threshold of the other world, and he is to be given the AAKHUT, or the sacred mushroom.

We now recall the description given by Harry Stone of such a door, and also his indication that certain rites (purification by water) are necessary in order to get through this door. At the same time, the Ra Ho Tep personality wrote in hieroglyphs the word RESTAU. Again we find a correlation between the events of A.D. 1955, and the beliefs of 2700 B.C. The identification of the King with Seker indicates where he is on his journey to the other world, and this early stage of the passage is here represented.

LINE 445c.   *N. is on the way to the place of Seker, chief of PḌUS.*

PḌUS is to be understood as a place similar to RESTAU as a domain of Seker as the god of the dead. There was also a people, and a land, of the same name, over which the King of Egypt ruled, but they have not yet been identified. The people are known as the Nine Bow People. The word PḌUS comes from the weapon—a bow, and is also used to describe the arch of the sky, as well as the divine Ennead. This latter association may indicate a celestial region PḌUS as the abode of the Ennead.

LINE 445d.   *This is your companion [other self?] who brings this [sacred mushroom] MaḌU for these of the wasteland.*

This is a difficult passage to interpret. We have seen in line 445b that the deceased King is identified with Seker, the god of the dead.

THE SACRED MUSHROOM   199

Reference to the companion may mean that the King, because of the identification with Seker, is now the other self as Seker. The King brings the offering of the golden mushroom, which is described as MaDU. The word MaDU is not known, and occurs only in this line. It is descriptive of the sacred mushroom as something needed by the inhabitants of the wasteland (the necropolis or limbo). I would suggest that this word be compared to the Vedic MAD, which we know as mead, or a divine ambrosia, descriptive of the sacred plant, the SOMA, or HAOMA.

The preceding four lines serve as an introduction to Utterance 301.

UTTERANCE 301

LINE 446a.   *To say: the ancient offering is thine O NUN together with NAUNET.*

The ancient offering is the AAKHUT, the sacred mushroom. NUN is the great primeval abyss out of which all arose. This abyss was personified as the god Nun, and his consort the goddess Naunet. It is interesting to note that the ancient Egyptians personified primeval matter as male and female, as positive and negative, as of dual composition. In the beginning was Nun, and out of this arose the primeval earth mound KA, and then the great god Atum appeared. Atum is the head of the nine great gods, or the Ennead. Since this utterance is an invocation, the first invocation is made to Nun and Naunet as the source of all things and gods created.

LINE 446b.   *Ye two sources of the gods, nourishing the gods with your SHU:*

The hieroglyph for sources is a jar or vessel, of the type usually determinative of the water god Knum. Thus it is a pair of vessels that are the sources of the gods, and these are Nun and Naunet, the dual sources of the manifested world. The hieroglyph for nourishing contains the same vessel (of Knum) as the word for sources; but in addition is coupled with the sign for SHU. This combination of the vessel of Knum and the SHU sign gives rise to the idea of

200   THE SACRED MUSHROOM

wellsprings, as the source of watery nourishment, and I have translated this as nourishing. This combination of source and nourishing
(denoted by Knum vessels) is followed by the word SHU, for which
I have included the original hieroglyphic writing with its mushroom
determinative. I have earlier discussed the word SHU written in this
form, and given my reasons why it should be interpreted as mushroom rather than the standard meaning of shade. I believe that there
is present in line 446b a subtle interplay on words around a root idea
of the mushroom as a jar or vessel, and this is personified as the
pair of gods Nun and Naunet as sources of the gods and watery
nourishment of the gods. See line 249 of the Pyramid Texts for the
same idea. (Chapter XIII, page 219.)

Nun and Naunet as personifications of the primeval abyss are here
presented as polar creative agents of this "nourishment." References
to such a pairing, or duality, in connection with the mushroom we
already know from Mexican sources, but the same references can be
found elsewhere in the Egyptian literature, and I present some selected examples:

In the Pyramid Texts, line 264, I give the standard translation by
Mercer:[2]

To say: O ye two contestants [the hieroglyphs for this phrase are:

announce now to the Honorable One in this his name.*
N. is this SŠSŠ plant which springs from the earth.

The phrase, ye two contestants, is derived from a pair of mushrooms, and these have been erroneously interpreted as opposing
weapons, hence the idea of contestants. The correct reading should
indicate pairing, or polarity, of two mushrooms, instead of contestants. This reading is confirmed by the line which follows, where the
SŠ plant is given in duplicated form, and clearly refers to the two
contestants (mushrooms) in the preceding line. The SŠ plant has
not yet been identified. SŠ (SESHU) must be descriptive of a
plant that springs from the earth. What could be more descriptive
of the rapidity of mushroomic growth than the term, springing?

In the Pyramid Texts we find other contexts where the word SHU
is shown in the dual or pairs. In line 1487a the word SHU appears

---

[2] *Ibid.* Vol. I, p. 76, Vol. II, p. 122.

THE SACRED MUSHROOM 201

with a determinative in the form of a mushroom with a double stalk, indicating a pair.

In line 1560, the deceased King is described ascending to heaven:[3]

> *He shall fly like a cloud to heaven, like a heron;*
> *he shall pass by the side-locks of the sky.*
> *The feathers [SHU] on the two arms of N. shall be*
> *like two knives.*

Feathers (which are also called SHU) are compared to two knives. The SHU is the means of ascent to heaven disguised as feathers whereon one travels.

In line 1090b it is stated that N. is lifted up on the wings (SHU) of the cow goddess Hathor. The agent for travel is again SHU, which has been translated as wings. This use of SHU as being the active agent that propels the soul to heaven is in harmony with the belief that the mushroom can cause the soul to travel outside of the body.

In line 1487a we again have SHU as a double-stalked mushroom where the deceased is called Osiris:[4]

> *Thou art standing, Osiris; thy SHU is over thee, Osiris.*

This usage of SHU recalls the vignette from the Book of the Dead showing the mushroom over the head of the deceased.

These few references illustrate my belief that the word SHU in such passages should be translated in terms of the mushroom ideology which we have been expounding.

LINE 446c. *The ancient offering is thine, O Amun together with Amunet.*

Amun is the god who later was to become identified with Ra as the One God. Amunet is the female consort of Amun. In the Pyramid Texts, Amun appears only three times, and is closely associated with the primeval gods Nun and Naunet. The ancient offering is, of course, the sacred mushroom. It is interesting to note that even in this early period of Egyptian history it was already an ancient and well-established offering.

[3] *Ibid.* Vol. I, p. 243.
[4] *Ibid.* Vol. I, p. 236.

202 THE SACRED MUSHROOM

LINE 446c. *The ancient offering is thine, O Amun together with Amunet.*

The repetition of this line suggests that the ancient offering was made in turn to the god and the goddess who personified the sacred plant. It is interesting to note that the HEB-SED ceremony was dedicated principally to two gods, Amun and Min.

LINE 447a. *Thy ancient offering is thine Atum together with the two lions [SHU and TEFNUT] doubly divine power, yourselves created of yourself.*
LINE 447b. *That is, SHU together with TEFNUT [who] created the gods, begat the gods, established the gods.*

The litany of invocation is here addressed to Atum, who is considered to be the first of the gods to appear out of the primeval watery abyss on the primeval earth mound, KA. In this passage he appears alone, although elsewhere his hand was personified and deified as his wife (line 1210b. AWSAAS) whereby he brought into being Shu and Tefnut. Shu is the god of air, and his consort Tefnut, the goddess of moisture. Shu as the god of air who holds up the sky must in some way be related to the Shu which gives lift in the ascension of the deceased on his way to heaven. This supposition is a personal interpretation on my part. The Texts, of course, do not associate Shu, the god, with the sacred mushroom, except as I have interpreted it.

This line ends the invocation made to seven gods, each of whom personifies the sacred ancient offering, the mushroom.

LINE 448a. *Say ye to you Father Ptah.*
LINE 448b. *That N. has given to you your most ancient offering, that N. has satisfied you with your due.*

The name of Ptah occurs only three times in the Pyramid Texts. Yet he is considered at times as a member of the Ennead, and held a most sacred position. PTAH means to open, and presumably the ceremony has approached the point where the way will be opened (for the supplicant). Ptah has been compared to the Ugaritic (a city in Syria to the north of Byblos) god KTR, the craftsman god. Here is another indication of an association of the mushroom ceremony

THE SACRED MUSHROOM   203

with the East, and particularly Syria. The ancient offering is presumably still the sacred mushroom.

LINE 448c.   *Ye shall not hinder N. when he ferries to the horizon to him [Ra as Atum].*

This line is highly reminiscent of the necessity in classical mythology of presenting a golden bough as a passport to the other world. The hero, Aeneas, in the following passage, has already plucked the golden bough from an oak, and in company with the prophetess has entered the passage to the underworld. *The Aeneid*, VI. 388.[5]

So therefore they [the prophetess and Aeneas] proceeded with the journey as before, and as they approached the river the Boatman, who, while still afloat on the Styx, had seen them in the distance walking through the silent wood and turning their steps towards the bank, spoke first and spoke in reproof:
"Whoever you are who stride in arms towards my river, come, say why you approach. Check your pace; speak now, from where you are. This is the land of Shades, of Sleep, and of Drowsy Night. It is sin to carry any who still lives on board the boat of Styx."

In reply Apollo's prophetess answered briefly:

"We have no such treacherous intent. These arms threaten no violence. Forego your alarm. Your monstrous guardian at the gate may fiercely howl in his den to all eternity, affrighting the ghosts till they turn pale. Proserpine may stay in fidelity behind her uncle's door. Trojan Aeneas, illustrious for his true righteousness and for his feats of arms, travels in quest of his father down to Erebos' deep shades. But if the sight of fidelity so strong has no power to move you, you must yet recognize this branch."
And she showed the branch (the golden bough) which had been hidden in her garment. The storm of anger in Charon's heart subsided and he said no more to them. He looked in awe at the holy offering, the Wand of Destiny, which it was long since he had seen. He turned the blue stern of the boat towards them and came near the bank. Next he hustled away the souls who sat side by side on the long benches, opened up the gangways, and immediately ad-

[5] Translation by W. F. Jackson Knight. Penguin, 1956.

204   THE SACRED MUSHROOM

mitted Aeneas, in all his bulk to the hull. Groaning under the weight, the stitched coracle let in much marsh-water through its leaks. They crossed the river; and Charon eventually disembarked both the priestess and the hero, unharmed, on ugly slime amid grey reeds.

LINE 449a.   N. knows him, knows his name. NḤI is his name.

It is Ptah who is still being addressed here, and who receives this word, NḤI, from the King as N. NḤI means the desired, or that which cannot be attained, the everlasting, the eternal. This word may also have served as a password, in addition to the ancient offering (the sacred mushroom). From the New Kingdom and on, NḤI is known as a sun god. At this point in the ceremony, it is known that the deceased is on his way to the sun god, Ra, who is in the Eastern Horizon.

LINE 449b.   He with the warrior's arms. Horus who is over the SHDU of heaven, who causes Ra to live every day.
LINE 450a.   He will rebuild N. He will cause N. to live every day.

This is the conclusion of the invocation to Ptah, and of the litany to the Ennead. These lines invoke titles of Ptah, warrior's arms, and Horus. The SHDU are the stars which were conceived of in this early period of sky worship as the abode of the souls of "the imperishable ones." Rebuild N. means to transform and to transfigure the King. This heralds the beginning of the ritual text.

THE RITUAL TEXT

LINE 450b.   N. approaches thee, Horus of ḤAT, N. approaches thee, Horus of ŠSMT.

A distinction must be made between Horus the Elder, the son of Atum; and Horus the Younger, the son of Isis and Osiris. Horus the Elder is the one referred to in this line. He was originally a heaven god, his two eyes being the sun and the moon. His symbol is the falcon, and he is represented as such. His most important forms are: HARACHTE, that is, Horus of the Horizon; Horus the Elder; and Horus of BHD.T. He was associated with the Nine Bow People which we have mentioned earlier.

Horus of ḤAT refers to a place, and its location is unknown. The

THE SACRED MUSHROOM 205

name ḤAT is pronounced KAT, and is of interest because the word is derived from the hieroglyph:

⊏

the Shu sign. In the Old Kingdom this sign was used to represent the SHU sound at times, and the KH sound at other times, as in this instance. We have already noted that the Pyramid Texts refer a number of times to a land to the east, or the northeast, of Egypt. This may point to Syria on the eastern end of the Mediterranean opposite the island of Cyprus. This land was called ḤAT, the home of the ancient Hittites; in the Indo-European ḤAT is KHATTI. The Egyptians knew this country well because it was the source of their cedar wood. The Egyptian word for cedar is:

ordinarily translated as, (Š; but it could also be, (Ḥ, pronounced as HAK. If this latter interpretation is admissible, then it may be possible to relate Horus' locale in the East, ḤAT, with the land of the Hittites, ḤAT, or KHATTI, and the Cedars which grew there (Š, or (Ḥ. The Lebanon where the Cedars grew is called ḤTAU by the Egyptians (*Wörterbuch* III, 349).

LINE 450C. *N. comes to thee, Horus of the East.*
LINE 451a. *Behold, N. brings to you his eye [soul] to be healed.*

In line 450b Horus of ŠSMT is identified as the sun god of the morning in line 342c. The locale of ŠSMT is unknown, and Egytologists have surmised that it is a land to the east of Egypt. Since this is a ritual text concerning the opening of the eye (by means of an unguent which will be described later) in order to give the soul its wings, or to spiritualize the King, we will probe a little deeper into the meaning of ŠSMT.

One of the meanings of ŠSMT is a flaming or fiery eye, comparable to the sun. The eye of Horus means the abode of the soul, and it is the aim of this ceremony to give it the wings. We find that the word, ŠST, is a winged creature. The word ŠSMU is a god of salves and oils. We have mentioned earlier that there is an alabaster vessel out of which the gods pour, or dispense, life. The word for alabaster is ŠS and in its shortest form is indicated by a hieroglyph in the form of a looped cord. It is known that the sign of initiation into the Greater

206 THE SACRED MUSHROOM

Mysteries of Egypt was a looped cord. Thus we find a group of Egyptian words built around the stem ŠS which may well apply to the mushroom ideology. On the basis of this analogy it is possible to freely translate the phrase, Horus of ŠSMT, from the point of view of a mushroom ritual as: Horus, the kindler of the eye [soul] who gives it wings and life. The King, as the supplicant in the initiatory rite presided over by Horus, is to be spiritualized and assume his rightful place among the gods.

LINE 451b.   *Take it, to thyself from N., kindle it! [Make it glow.] Its water is in it, kindle it! [Illuminate it.].*

The word for kindle has as its main hieroglyph a picture of a fire drill as the instrument whereby fire is generated. Actually, this word in the Pyramid Texts is not to be found in the *Wörterbuch*, and I have given it this meaning of kindle on the basis of the context of the stage reached in the initiatory ritual, and the use of the fire-drill sign.

The water in the eye does not mean ordinary tears, but a suffusion with the great water from heaven as described earlier by the Ra Ho Tep personality, and from the HEB-SED ceremony. This may be considered as a symbol of godly blessing, as well as purification.

LINE 451c.   *Its blood is in it, kindle it! Its air [pneuma] is in it, kindle it!*

The King's soul is offered for transformation under the symbolism of fire, water, blood, and air. Each of these is to be spiritualized by the grace of Horus. Fire, of course, is the symbol for the activation of the other three elements. This can best be understood in the light of the discourse I have already given on wood, and its resident spirit.

Blood, in ancient Egyptian thought, symbolized victory over life in the same way that I have already described the red water from the inscription in the tomb of the historical Ra Ho Tep in connection with AKHUT. The blood sacrifice is well known as one of the most time honored of offerings to the gods in many cultures. The kindling of the blood symbolizes the final victory of the soul over the mortal life of the flesh.

The hieroglyphs which I have translated by air are interesting, and are given by a set of pictures which mean throat, or voice. This same set of hieroglyphs when conjoined with

SRK, means to open the throat to get air. The determinative, in the shape of a mushroom on its side, is silent, and the *Wörterbuch* describes it as an unknown object. The presence of this latter sign in this word, SRK, is interesting in view of the importance which the Egyptians attached to the ceremony of the opening of the mouth, whereby the deceased is spiritualized by a symbolic opening of the mouth. SRK also means to open a door; to open the way; the sun rays; an emission of the sacred eye; the flesh of Horus; as well as some unidentified plant. This set of meanings which are attached to SRK forms a cluster highly suggestive of the sacred-mushroom ideology.

That SRK may really be associated with a mushroom rite gets added support from the following. On a VIth Dynasty coffin from Saqqara (Quibell, Volume II, Plate 2) I found the following sign:

The hieroglyphs used to explain this sign are ordinarily transcribed as KRS, meaning he who is in his grave. If these hieroglyphs are read in the reverse order, they spell the word SRK. If this be the correct reading, in association with this mushroom-shaped sign, then I believe that the word SRK may yet be another clue as to the nature of the hypothetical mushroom rite in ancient Egypt.

LINE 452a. *Ascend in it [the winged eye]. Take possession of it, in this thy sacred name of ḤKAS.*

There is a subtle play on words here between ascend as AAK and the sacred name ḤKAS. The verb AAK, ascend, is based on the noun for a ladder, AAK. This sign for ladder is the same as the one used by the Ra Ho Tep personality in the word for a sacred mushroom, AAKHUT. It may be that the ascension (ladder) is personi-

208 THE SACRED MUSHROOM

fied by the sacred name of ḤḲAŠ. The meaning of ḤḲAŠ is not known, but the hieroglyphs for it,

contain the sign:

The meaning of this latter sign is also unknown. But we already know this sign in its modified form (with the Ra, or mouth-shaped sign, around it) from the word for the sacred mushroom given by the Ra Ho Tep personality, as AAKHUT

In this word there is present the ladder sign as part of the sign for the mushroom. Thus the play on words here between ascend, AAK, and ḤḲAŠ becomes very subtle indeed.

This subtlety can be clarified somewhat by the occurrence of the word ḤḲAŠ in another passage (*Bulletin de l'Institute de Archeologie Orientale du Caire*, Le Caire, 45: 175–93) where it is found in association with two rare old divinities, HEPUI, and ḤDD, of the marshes of the Delta. In this passage, ḤḲAŠ, has been interpreted as a covering, or clothing, which is deified. This is all that is known. However, the name which has been interpreted as HEPUI comes from the following hieroglyphs:

These show a door sign with a mushroom-shaped figure on each side. The presence of these signs, as a pair, one on each side of the door, is most intriguing. Since line 452a beckons the deceased to ascend, or enter, a new state, we can assume that a passage through a portal, or a door is both symbolic and appropriate to this stage of the ritual. Apparently, ḤḲAŠ is symbolic of the assumption of this new state of being in the presence of Horus.

We find in a passage from the Pyramid Texts, Utterance 305: "To say: the ladder is fastened by Ra in the presence of Osiris; the ladder is fastened by Horus in the presence of his father Osiris as he (the deceased King) goes to his AḤ [AK]." Here there is a play on the

THE SACRED MUSHROOM   209

words ladder, AAK; and the spirit, AH. In our line 452a, the deceased
is also on his way to transformation as an AH, a spirit. This play on
words is further illuminated by a consideration of the first sign in the
hieroglyphic for ḤKAŠ, the scepter, or ḤKA. ḤKA, as the scepter
sign by itself, means a prince or a ruler. This meaning would serve
to emphasize the power of the new state to which the deceased
has ascended. ḤKA also means magic, as a passenger in the sun boat.
This refers to the magical transforming power of the process whereby
one ascends to heaven, in this case symbolized by the sun boat. I believe that the associations I have cited in relation to the word ḤKAŠ
all serve to unravel its obscurity, and place it in relation to the practices and ideas which are part of the sacred-mushroom ideology.

LINE 452b.   *Thou ascendest to it in this thy name of Ra.*

It is quite apparent that the ascension and transformation of the
deceased King has now brought him into identification, if not
equality, with Ra. But that the transformation process is not yet
complete is indicated by the following line.

LINE 453a.   *Put it on the top of thy head [aperture of Brahma] in
this its name of unguent [for the top of the head].*

This difficult line can only be clarified when we recall that the Ra
Ho Tep personality described on June 16, 1954, the application of a
mushroom unguent to the skull sutures at the top of the head. The
Buddhists call this site the Aperture of Brahma, and believe it to be
the portal of exit of the soul from the body. The combination of
information from the Ra Ho Tep personality, the Buddhist tradition,
and the place of this line in the context of the Egyptian ritual, serves
to unlock the enigma.

We know from the previous line that the deceased is about to un-
dergo transfiguration as a spirit. He is now anointed as a sign of such
accession. The word for the top of the head, HAT, and unguent,
HATT (for the top of the head), is again a play on words. To this
play on words we can relate the rare word for "the weak-place on
the top of the head," AHT, as signifying the Aperture of Brahma.
In the light of our mushroom ideology this is no accidental relationship—the words HAT, AHT, and HATT are clearly related. By
means of the unguent HATT, the Aperture of Brahma AHT, on

the top of the head HAT is to be opened and the soul released from mortal bonds.

This is the first of four appellations given to the sacred unguent, and because it is to be applied to the Aperture of Brahma, I choose to designate HATT as the "Opener."

LINE 453b.   *That thou may rejoice in it, in this its name of TRT tree.*

The word for rejoice is TRRU, and it is clearly a play on the name of the tree, TRT. This play on words leads me to believe that the TRT tree must in some way be connected with the unguent applied to the top of the head. It may well be that the HATT is a mushroom unguent, and that this latter is some other unguent. It is not known what species of tree the TRT is, although Egyptologists have interpreted it to be a willow tree. It is known that the TRT was a sacred tree; that there was a sacred ceremony of setting up a TRT tree (or pillar); that the TRT, or some part of it, was used in medical prescriptions for open infected wounds; that the TRT tree was deified as a god; and that the TRT is an unknown something that the Horus eye (soul) came into possession of.

In analyzing this problem from the sacred-mushroom knowledge at our disposal we can make the assumption that: (1) the tree TRT could have been associated with the growth of a mushroom (as is the oak, birch, etc.), or (2) that the oil in the unguent came from this tree. The reader may recall that on September 4, 1954, I tried to find out from the Ra Ho Tep personality what tree or plant was the source of the oil used to make the mushroom unguent. I got no clear answer to this question, except that Harry enacted an answer to my question by vigorously rubbing, or grinding, on the table. I interpreted this response to mean that he was grinding something between his hand and the table, and he indicated that my surmise was correct. The reader can well imagine my surprise, therefore, when three years later I found a hieroglyphic reading for TRT in the *Wörterbuch* as follows:

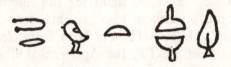

THE SACRED MUSHROOM   211

The *Wörterbuch* translates this word, TRT, as a willow. The meaning of the two opposed cup signs is unknown. In the light of our mushroom preparation insight, it is possible that these two objects represent two grinding blocks, presumably of the TRT wood. Some form of grinding block or mortar must have been used, if we are to place any credence on the Ra Ho Tep personality statement, either to grind the mushroom, or to prepare the oil used in the unguent (from fruit, nuts, leaves, or bark). Whatever their function, they would, if used in the sacred-mushroom ceremony, be regarded as important cult objects. We could more clearly understand the role of the TRT tree in this part of the ceremony if we knew the species of tree that it represents, and therefore its products and pharmacological properties.

There has been recorded a medical preparation made from the TRT tree, called DRD. It is not known what product of the tree DRD represents. In the Berlin Medical Papyrus this product is found only once, and here it is called the "divine DRD." Again the meaning of this phrase is unknown. DRD sounds very much like the Greek word for the oak, DRD, from which came the name of the Celtic oak-worshiping people, the Druids.

This second appellation for the sacred unguent, now likened to the TRT tree, can be given a name from its medical uses as "that which brings or draws out."

LINE 454a.   *That thou may shine thereby among the gods, in this its name of that which illuminates [or THNUT oil].*

This is the third appellation given for the sacred unguent, and means that which illuminates. Again there is a marked play on words between shine, THNHN, and the word THNUT, which means illuminate or serenity, and is depicted in the hieroglyphic as an oil or unguent. Curiously enough, this same oil is present in the offering list of the tomb of the historical Ra Ho Tep, but here it has for a determinative the same form of jar or vessel that is used in the sign for KNUM, the god who in the beginning of Utterance 300 fashions the AAKHUT. This oil is also prominently listed with the important cult objects of the HEB-SED ceremony. This association may serve to identify the THNUT oil more closely with a sacred-mushroom rite than the two preceding appellations, and we can call it "the illumination."

212   THE SACRED MUSHROOM

LINE 454b.   *That thou may be blessed with it in this its name "praise to the highest" oil [HKNU oil].*

The play on words here is most intricate and subtle. The word for blessed, HKN, is paired against the word HKNU, which generally means "praise to the highest," i.e., to the godhead. With this phrase the deceased King has reached the height of spiritualization, and is compared to, or identified with divinity and the godhead. There is undoubtedly here a further play on the word HK, as magic personified, and NU, which means waters (as great waters). When magic is personified it is ascribed to the godhead as a prime attribute; and when HK is coupled with NU, waters, it would mean the manifestation of magic. In short, the transforming magic of the great waters from the godhead.

HKNU is also the name of a sacred oil, and it, too, is found in the offering lists in the historical Ra Ho Tep's tomb. HKNU is also a place name in Syria that has not been definitely identified but is believed to be in the region of Byblos. This region of Syria, HKNU, was ruled over by an ox god (not known to be related to the ox of the tree deity Yayash mentioned earlier).

The fourth appellation of the sacred unguent, HKNU, may be designated as "divine power," as descriptive of this stage of the ritual of transformation.

LINE 454c.   *RNN.UT.T now loves thee.*

RNN.UT.T means linen, or linen weaving, personified as a goddess. She is also the harvest goddess, goddess of weaving, goddess of magic, and the uraeus serpent on the brow of the sun god. In these many forms of the goddess we catch a glimpse of the divine attributes of the now transformed and transfigured soul of the King. The uraeus serpent symbolizes eternity.

This line concludes the ritual portion of this passage. The remainder of the passage is addressed as a litany to the spiritualized King under the form of the god Sept. Sept is also known to be prominent in the HEB-SED ceremony. His attributes are praised in lines 455 and 456, which we will pass over and turn our attention to the concluding lines of the litany.

LINE 457a.   N. is purified. N. is an AHKHUT.
LINE 457b.   In SHU, this, thine. Jackal, O Jackal, where thou purifiest the gods!
LINE 457c.   Thou art become a soul,
Thou art become Sept,
Horus, lord of the green stone
[like] the two green falcons.

The transfiguration is complete. The soul of the deceased has been elevated to that highest degree, the AHKHUT, now depicted with a unique set of hieroglyphs:

This word AHKHUT is now symbolized by a bird—sign of the imperishable spirit; the primeval hill—the first creation of the Creator; and a tree—domicile of the spirit on earth. The King's soul is now equal to, and in the company of, the god Shu—aerial spirit. The jackal—symbol of guardianship of the secret of soul mysteries, the hermetic vessel of transformation—belongs to the King, too. He is ranked in the company of Sept and Horus the Elder. Upon him have been conferred the symbols of the green stone and the green falcons. Such is the imagery and symbolism which the ancient Egyptians clothed around the arcana of the soul. Their key to the door of eternity is AAKHUT, a golden talisman to the beyond.

APPENDIX 4

Margaret Murray[1] has brought to light a remarkably widespread pre-Christian religion extending from India to Europe, from Paleolithic times down to our seventeenth century. It is widely known in Europe in connection with the persecution of witches, and its central figure is known as the Horned God. Dr. Murray has shown that the earliest record of the existence of this god is to be found in the Paleolithic cave paintings of France. The usual scene is that of a man clothed in the skin of a horned animal and dancing out some unknown ritual. She has shown that this Horned God was sacred to both the hunting and the pastoral folk of early times. She has traced this cult of a Horned God in the Bronze Age of Egypt, Mesopotamia, India, and Greece.

In reference to the Horned God, she states: "In Egypt his horns were those of Amon, the supreme god. In Egypt—with the exception of Mentu—the horns of cattle are worn by goddesses only, while the gods have horns of sheep. The chief of the horned gods of Egypt was Amon, originally the local deity of Thebes, later the supreme god of the whole country. He is usually represented in human form wearing the curved horns of the Theban ram. Herodotus mentions that at the great annual festival at Thebes the figure of Amon was wrapped in a ram's skin, evidently in the same way that the dancing god of Ariège (the Paleolithic portrayal) was wrapped. There were two types of sheep whose horns were the insignia of divinity; the

[1] Margaret Murray, *The God of the Witches*, reprinted by permission of Oxford University Press, Inc., New York, 1953.

Theban breed had curved horns, but the ordinary breed of ancient Egyptian sheep had twisted horizontal horns. The horizontal horns are those most commonly worn by Egyptian gods. One of the most important of these deities is Knum, the god of the district around the first cataract; he was a creator god and was represented as a human being with a sheep's head and horizontal horns.[2]

"There are two other links between Egypt and the dancing god of Ariège. On a slate palette, which is dated to the period just before the beginning of Egyptian history, there is represented a man with the head and tail of a jackal (Quibell, J. E., *Hierakonpolis*, II, Plate xxviii, Ed. 1902); as in the Ariège example, the body, hands, and feet are human; he plays on a flute, and like the Paleolithic god he is in the midst of animals. The other link is in the ceremonial dress of the Pharaoh, who on great occasions wore a bull's tail attached to his girdle. The SED-HEB or Tail Festival, when the King was invested with the tail, was one of the most important of the royal ceremonies. A sacred dance, performed by the Pharaoh wearing the bull's tail, is often represented as taking place in a temple before Min, the god of human generation."[3]

These links between the Paleolithic Horned God and the Egyptian HEB-SED (the ceremony is called either the SED-HEB, or the HEB-SED) discovered by Dr. Murray are of the greatest interest. I have already pointed out my reasons for believing that one of the aspects of the HEB-SED ceremony was a sacred-mushroom rite. I have studied in the same Paleolithic culture where Dr. Murray reveals the Horned God an interesting carving found on a section of a reindeer horn.[4]

The curious fact about these three symbols is that each of them is prominently displayed in the portrayals of the Egyptian HEB-SED ceremonies thirty millennia later. The first figure, in the shape of an A, is prominently displayed in the hands of a prehistoric Pharaoh

---

[2] *Ibid.*, p. 25.
[3] *Ibid.*, p. 26.
[4] *Scripta Minoa*, Sir Arthur Evans, Oxford, 1909. Fig. 2. Groups of such signs are seen in sections of reindeer horn from the Grotte de Lorthet, Hautes-Pyrénées, dated from Mousterian times (Middle Paleolithic).

THE SACRED MUSHROOM   217

as a hoe (Egyptian MER) in the form of an A.[5] Wainwright[6] identifies this scene as a HEB-SED ceremony. The Pharaoh is portrayed in historic times from many other scenes carrying this same hoe in HEB-SED ceremonies, and it has been interpreted by some authorities as being concerned with land and fertility rites.

The second sign in our Paleolithic carving is that of a U-shaped or semi-ellipsoid figure. This sign, too, is portrayed prominently in many HEB-SED ceremonies, particularly in the stela of King Zoser which I have portrayed on page 158, showing the pair of mushrooms behind the King, although I have not included the signs presently under discussion. It is not known what these signs represent with any certainty, except that they are always shown in two groups of three each.

The third sign in our Paleolithic carving is definitely of a mushroom shape. Its presence in the HEB-SED ceremony is to be seen in the figure on page 158. The duplication of these three signs in the Paleolithic carving and the Egyptian HEB-SED ceremonial portrayals adds another link to those already discovered by Dr. Murray.

It is a further curious coincidence that the geographical area covered by the cult of the Horned God is the same area as is covered by megalithic monuments in the form of mushroom-shaped stone slabs, or cromlechs. It appears to me that these findings are of sufficient interest to encourage archaeologists and anthropologists to reexamine the data from this area and its various historical periods.

Although a great deal has been written about the HEB-SED ceremony, I have yet to find a single authority who has even suggested that it might be a sacred-mushroom rite. I would like to raise this question, and speculate on some of the reasons which lead me to identify it with the sacred-mushroom practice. Wainwright[7] is of the opinion that the HEB-SED is part of the prehistoric sky and fertility religion, and that as portrayed it is essentially a running ceremony for the King. He is further of the opinion[8] that this ceremony was concerned with the sacrifice of the King. By this he refers to the thesis so ably advanced by Sir James Frazer that kingship existed by virtue of the magical powers that a given individual possessed. By

[5] *Primitive Art in Egypt*, Jean Capart, Lippincott, London, 1905. Figs. 188 and 189.
[6] *The Sky Religion in Egypt*, G. A. Wainwright, Cambridge, 1938, p. 18.
[7] *Ibid.*, p. 18.
[8] *Ibid.*, p. 86.

218   THE SACRED MUSHROOM

such magical powers the King was able to influence the fall of rain, the fertility of land and beast, and by divination ascertain the will of the gods. The whole community depended on his skill and powers. Should these powers fail, the well-being of the community was in danger. Hence arose the curious practice of periodically sacrificing the King while he was in his prime, so that he in turn could pass on his power intact to his successor. At some period in history men came to the realization that since the King had power by virtue of magic, he could also restore such waning powers by means of magic. One of the means used to escape sacrifice was to single out a victim who would be made a mock King for a day and then give his life as a substitute for that of the King.

However, we know that in the case of the HEB-SED ceremony, as the King got older he repeated this ceremony more frequently, and some have interpreted this to mean that the purpose of the rite was to secure rejuvenation of his powers. There is much justification for this thesis. Let us make the assumption that one of the purposes of this ceremony was indeed rejuvenation, and analyze its application to facts that we now have in hand about both the sacred-mushroom rite and the HEB-SED ceremony.

Let us recall that one of the main functions of the sacred-mushroom ritual is to confer upon the user certain powers which we deem to be of a magical nature. These include the power of extra-sensory perception in finding lost objects, foretelling the future, and powers of healing the sick. In addition there is the Siberian shamanistic tradition of supernormal strength, endurance, and the travel of the soul to the realm of the spirits and the gods. These are, in fact, the same powers that earned for one the right of kingship in early times. Therefore, the mushroom, if known, could be used for the rejuvenation of the King's waning magical powers. But since the custom of the sacrifice of the King was deeply ingrained in the community it would require more than a show of power on the part of the King to escape death. It would require the consent of the tribal gods in order for the King to continue his office beyond the appointed term.

To this speculation we must couple the widespread belief in the dying god who is resurrected, i.e., Dionysos, Osiris, etc. It seems that the aspiring King, in order to prove his right to continued kingship, would also have to die and in a sense be reborn. It appears most probable that the ancients were thoroughly familiar with a wide vari-

THE SACRED MUSHROOM   219

ety of natural poisons. This would be especially true of the *Amanita muscaria* in the light of the traditional poisonous nature ascribed to it. The ancients must have been just as puzzled, as are modern doctors and mycologists, at the fact that some people who eat liberally of this plant do not die as is expected. Instead they report, no doubt in glowing detail, all of the wonderful hallucinations that they had had while in an apparently dead state, but actually in an inebriated stupor. Just as there are trials by fire, and other lethal ordeals, so some enterprising clan priest must have hit upon the idea of using the mushroom as an ordeal of death. If the unhappy victim survived he would only add more information to the clan archives about this mysterious mushroom. It is possible to conjecture that some individual who was subjected to such an ordeal was of a sensitive nature, i.e., as I have described for Stone and Hurkos. His reaction to the mushroom would be extraordinary and his experiences so fabulous that they would be ascribed to the agency of the gods, as they are in Central America today. Such arcana would be seized upon by the tribal elders and the shaman and classified as a deep dark secret to be handed on only to the initiates. And as clan custom is now known, this would be handed down the priestly line as a part of the stock in trade in magic.

Now the continuance of such a custom is vitally dependent on having access to a fresh supply of *Amanita muscaria*. Fortunately this plant stores well and an advance supply could be maintained for years. If the tribe moved, or was driven out, the source of supply would be cut off; or be accessible only upon long journeys to the fountainhead. Now in the case of the Egyptians, one of the reasons that no one has suspected the presence of a sacred-mushroom cult, is the absence in the country itself of the vital plant. However, there are many reasons for believing that some of the early Egyptians either came from the Syrian forest area or were thoroughly familiar with it. It is my thesis that the Lebanon Mountains and the Amanus Mountains were the source of the mushrooms used in the HEB-SED ceremony. When this practice started or died out in Egypt we do not know. All we know is that the description of the mushroom in the texts and ritual representations cease with the age of the Pyramid Texts, the VIth Dynasty. In any case, by the presence of the elements that I have mentioned, there is good reason to assume the existence of a sacred-mushroom cult in ancient Egypt. And the avail-

220 THE SACRED MUSHROOM

able evidence suggests that its usage occurred in the HEB-SED ceremony.

Since it is known that the ancient Egyptians in prehistoric times sacrificed the King, we ask the question as to how they came to give up this practice. This practice would only be given up if a suitable substitute were found, namely, one in which the King either maintained or rejuvenated his magical powers. It is my contention that a sacred-mushroom rite would fulfill this condition. By using a drug, such as the *Amanita muscaria*, the prime question would be settled by ordeal; either the King survived or he did not. Hence, this practice is tantamount to a trial by poisoning. But we cannot doubt that the priesthood through long experience had learned how to use this drug with safety, even as we have had the safe use of this drug explained by the Ra Ho Tep personality.

If the King did survive, he had to prove his prowess in a number of ways. First, he had to show his physical strength and endurance during the ordeal of running and dancing portrayed in the HEB-SED ceremony. This phase is highly reminiscent of what we know of the shamanistic practice. Second, like the shaman, the King undoubtedly had to traverse the underworld, probably in the manner prescribed in the Book of the Dead. Here he had to meet the forty-two gods of the underworld and secure the permission of each one to continue his earthly reign. In this conjectured phase of the ceremony the HEB-SED portrayals show the King accompanied by a special priest. We can only surmise that the priest was there, both to help the ordeal of the King, and also to verify the underworld journey of the King, should he be successful. In this manner we surmise that the King was able to escape sacrifice and to continue his reign until the next HEB-SED.

The interesting question is raised as to why the IIIrd Dynasty King Zoser is the last to portray the HEB-SED ceremony with a display of the sacred mushroom. The Pyramid Texts displayed after this period give only the barest hints of such a rite. And why should Ra Ho Tep, as manifested through Harry Stone, be privy to such sacred secrets? There is a plausible answer to these questions.

My studies have shown that the fragmentary material known from ancient Egypt up to the time of King Zoser does not attempt to hide the representations of the sacred mushroom. It is known that the reign of Zoser marked a turning point in Egyptian culture. For one thing it is the visible beginning of the pyramid construction

THE SACRED MUSHROOM 221

era. In another respect certain cultural patterns were laid down for architecture and medicine during the reign of Zoser. The vizier of King Zoser, Imhotep, is usually given the credit by historians and by Egyptian tradition for these innovations. As such he was responsible for the architecture of the Step Pyramid; he is called the Father of Medicine; and he undoubtedly fulfilled the function of a high priest in his role as adviser to Zoser. These facts indicate that great changes occurred at just this period of Zoser and Imhotep, and the secret nature of the mushroom rite in the HEB-SED ceremony may well have been instituted at this time.

The historical Ra Ho Tep, as the great seer of Heliopolis, one of the first great prophets mentioned in Egyptian history, would of course be privy to such secrets. But there is another link of Ra Ho Tep with Imhotep which is given by J. B. Hurry in his work on Imhotep.[9] Here it is stated that a long genealogy was discovered on a stela listing a long succession of architects from the Old Kingdom down into late history. This list makes Ra Ho Tep the direct descendant of Imhotep. Although this is possible in terms of the number of years intervening between the two figures, it is difficult to reconcile with the other listings of Ra Ho Tep as the son of King Snefru. It is further curious that both Ra Ho Tep and Khufu are variously listed as the sons of King Snefru. The former became the high priest at Heliopolis, while the latter became the mighty Pharaoh of the Great Pyramid. Since this question cannot be resolved, I can only remind the reader that the Ra Ho Tep personality stated on June 16, 1954 (page 15), that he was brought up by an architect. There is another bit of information which I have not previously included in this book, and that is that the Ra Ho Tep personality has stated that he met his death by murder at the hands of the one who assumed power, and that this one who assumed power (he does not give any name) then closed or destroyed all the temples with which Ra Ho Tep had been familiar. I only add this information as a part of my reporting. We do know historically that Herodotus states that the Pharaoh Rhampsinitis (now known to be identified with Khufu) closed all the temples and placed the land under forceful rule. It is also known historically that this period coincided with the triumph of the priests of Ra who firmly established their god as the supreme ruler of Egypt.

The above considerations, facts, and conjectures, of course, do not

[9] *Imhotep*, J. B. Hurry, Oxford, 1928. This genealogy is reproduced by Hurry from *A History of Egypt Under the Pharaohs*, II, H. Brugsch-Bey, 1879, p. 299.

222   THE SACRED MUSHROOM

pretend to offer any definitive answer as to why the sacred-mushroom rite was lost to historical view at this period of Egyptian history. For all we know the rite may have been practiced as long as the history of the HEB-SED ceremony, but in secret. On the other hand, there may have been a loss of continuity in the tradition and practice during one of the periods of invasion or foreign occupation. It is apparent that much remains to be done to settle these historical questions, and especially in regard to the rite itself.

INDEX

Page numbers in *italics* refer to illustrations.

AAK, 207

AAKHUT, 139–40, 157–59, 184, 190–92, 196–200, 207–8, 213
defined, 131

AAKU, 183

Aeneas, 131, 203–4

Aeneid, The, 131, 203–4

AG, search for meaning of its root, 144–46, 152

AḤ, 154

AḤKHUT, 213

AHT, 154

alabaster dagger, 74–75

Amanita muscaria, 2–5. *See also* mushrooms; sacred mushroom/ sacred mushrooms
bad taste and smell while using, 98–99
chemical makeup of, 95–97
consuming safely, 96
described, 22
and disturbance of vision, 98
effects of, 22–23, 95, 97–99, 105, 164–66

and ESP, 99–100, 105
evaluating poisonous effects of, 97–98
failure to sprout in lab conditions, 90
finding—first sample in Maine, 82–84
finding—July 20, 1955, 87–88
finding—July 24, 1955, 88–89
finding—July 27, 1955, 89–90
finding—August 26, 1955, 91–92
finding—September 1, 1955, 92–93
and hot skin, 98
and hypersensitivity, 98–99
looking for / finding more samples, 86–94
and lowering of pulse rate, 98
and Maine-Mexico ESP experiment, 71–73, 85–86
mentioned in the first Stone session, 17–18, 21–22
in mystery traditions, 4–5
poisonous nature of? 22, 96–98, 219

223

224 INDEX

and soul flight, 2
Syriac medicine and, 144–46
testing first sample, 83
three problems to explore, 95
two colors of, 22
use in Siberia, 23–24, 142
uses of, 24
Amon, 215–16
Amun, 201–2
ANKH, 3, 29–30, 192
defined, 29
ANKH KHUT, as plant of life, 31
Antinea, 19, 45–46, 128
Anubis, 129
Aperture of Brahma, 153–54,
209–10
ASH NUAH, 76–77, 135
atropine, 96
effects of, 96

beetle sign, 189
belladonna, 96
Bernstein, Morey, 9
best stories, 9
block experiment, 63–66, 100, 176
defined, 63
blood, symbolism of, 206
Bouverie, Alice, 56–57, 162
asked to be cautious, 37–38
background of, 10
death of, 115–16
first session with Harry Stone,
1–2, 10–11, 17–20
and further early meetings with
Harry, 40–49

no remembrance of Egyptian
lifetime, 50
trance writings and speakings of,
July 4, 1955, 78–81
Budge, Wallis, 10
bufotenin, 96–97
effects of, 97

cartouche. *See* gold pendant
CHATTRA, etymology
investigated, 149–52
Cheops. *See* Khufu
Chinese culture, mushrooms in,
141–42, *141*
clairvoyance, defined, 62–63
clairvoyance explanation for
drawings, 170
"collective unconscious" explanation
for drawings, 170
consciousness independent of body,
20–21
Cromlechs, 158
crown of the head, 153–54, *153*
curandero, 134, 155
defined, 58
in ESP test, 72–73

deadly nightshade, 96
death at Permanente Hospital,
115–16
Dodds, E. R., 20–21
door to eternity, 198
drawings by Harry Stone. *See*
hieroglyphic drawings by
Harry Stone

DRD, 211
Druids, sacred plant of, 74–75
dying god who is resurrected, 218–19
dynasty, defined, 33

earth-ball prediction, 103
Egyptian language, what it sounded
    like, 124
Egyptologist, comments on Harry's
    drawings, 25–35
EN KATU, 3, 29–30, 127
    defined, 32
Ennead, defined, 130
ESP (extrasensory perception),
    38–40, 71–73
    author's report on, 12
    problems with research on, 15
ESP tests, 42, 63–66, 175–78
    Maine-Mexico, 71–73
"Eternity is watching," 137
EUL, 141
explanations for drawings / statements
    clairvoyance, 170
    "collective unconscious," 170
    mediumship, 169–70
    reincarnation, 162–63, 169–70
    telepathy, 27, 161–62, 170
extrasensory perception. See ESP
    (extrasensory perception)
eye of Horus, 205

Faraday Cage, 12, 175–78
    described, 175
    and ESP tests, 175–78
ferryman, 195–96

fly agaric. See Amanita muscaria
Frankfort, H., 156
Frazer, James, 131, 217

Gallow, Mr., telepathy experiments
    on, 62, 66–68
Gardiner, Alan, 123
Garrett, Mrs., 52–53, 55
golden bough, 131
golden plant, 74, 135, 186
gold headband, 132–33
gold pendant, 1–2, 10
    Harry's description of his
        experience with, 42–43
Great Pyramid, 33–34
Greeks, ancient, 49–50
GU, etymology investigated, 149–52

HAB, 69
HANGGO, 142–43
Harry. See Stone, Harry
HATT, 209–10
HEB-SED ceremony, 135, 165–66,
    211, 216–22
    mushrooms in, 159–60
    source of mushrooms in, 219–20
Heliopolis, 125, 130
Herodotus, 221
Hesse, Erich, 23
hieroglyphic drawings by Harry
    Stone, 113–14, 123. See also
    trance statements by Harry Stone
    first session, June 16, 1954, 10–11,
        17–20, 18, 179–80
    June 23, 1955, 134

226 INDEX

June 24, 1955, 134
June 30, 1955, 73–77
August 25, 1954, 182–84
August 29, 1954, 187
September 4, 1954, 181–82, 189–92
October 5, 1954, 187–89
December 9, 1954, *113*, 192–94
September 8, 1955, 137
analysis of, 123–28
authenticity questions, 25
dating of, 123–24
detailed interpretations, 179–94,
    179–94
explanations for, 169–70
meaning behind, 169–72
reviewed by Egyptologist, 25–35
summarized, 130
hieroglyphics, Egyptian, 25
number of, 123
hieroglyphics, miscellaneous images
    of traditional, *199–200, 205,*
    207–8, 210, 213, 216
ḤḰAŠ, 208–9
HKNU, 212
Horned God, 215–17, *216*
Horus, 204–5, 206
Hrozný, 147
Hrti, 195–96, 197
Hupe, 77
Hurkos, Peter, 100, 178
    and *Amanita muscaria*, 99–105
    described, 110–11
    drawing by, *114*
    further experiments in Maine,
        110–16

prediction for September 27, 1957,
    103–4
psychometry of, 110–11
sees large luminous mass, 112–13,
    115
Hurry, J. B., 221
Huxley, Aldous, 99–100
hypnosis experiments, 117–20

Imhotep, 221
immortality, plant of, 31, 34, 141

Jochems, Mr., experiments with,
    117–18

KATU, 28
KHEPER, 132
Khufu, 221
    described, 32
kings
    magical powers of, 217–18
    sacrifice of, 220
KM, 185
Knum, 157–58
Koryaks, and use of *Amanita*
    muscaria, 142

LING CHIH, 141, 146
Lord of Life, 46

Maine-Mexico ESP experiment,
    71–73, 85–86
MAT (matching abacus test),
    63–66, 100, 176
MCC, 166-69

INDEX 227

mediumship, 49

mediumship explanation for drawings, 169–70

MEDU, 124

Mexican mushroom connections, 57–59, 163

Mexico-Maine ESP experiment, 71–73, 85–86

mobile center of consciousness (MCC), 166–69

Mohawk Trail, 91–92, 94

Murphy, Bridey, 9n2

Murray, Margaret, 215–17

muscarine, described, 95

mushroom poisoning, 21–22

mushrooms. *See also Amanita muscaria*; sacred mushroom/ sacred mushrooms
effects of, 164–66
in Egyptian language, 139
ritual use of, 57–59

NA HA HE HUPE, 77

Nanacatl, 23

Nefert, 26

Newman, P. D., biography, 5

nitrous oxide experiences, 55

"noble waters from heaven," 74

Nolton, Colonel, 12, 15, 38

NUAH, 76–77, 135

NU AH A HADI, 77

opening the door, 47, 131, 137

oracles, 49–50

Osiris, 201

out-of-body experiences/dreams, 51–56
nitrous oxide experiences, 55

parasympathetic nervous system, 95

PAR UP KA, 118–19

PDUS, 198

Petrie, W. M. Flinders, 124–26

plant of life, 31, 34, 141

Pleydell-Bouverie, Ava Alice. *See* Bouverie, Alice

PR, 154

predictions, as approximations, 104–5

psilocybin, 105n2

psychedelic drugs, 3. *See also Amanita muscaria*; sacred mushroom/sacred mushrooms

psychometrist
defined, 64
Peter Hurkos as, 110–11

Ptah, 202–4

Ptah Katu, 47–48, 124, 128

PTAH KHUFU, 32

Puharich, Andrija (author). *See also specific events*
Army career of, 10, 12–16, 60
Army research on ESP, 12–15, 38–40
biography, 231
first call from Alice Bouverie, 10–11
first meeting with R. G. Wasson, 57–59
four stages of relationship with Harry Stone, 107–9

228   INDEX

offers Harry Stone a job, 60
out-of-body experience of, 51–56
returns to Maine, 1955, 61
security clearance of, 39–40
summary of his explorations, 1–5
visits Mrs. Bouverie in an OOBE/
dream, 53–55
Puharich, Jinny, 16, 61
Puharich, Lanie, 87–88
Pyramid Texts, 133–35, 146
sacred mushroom rites in, 195–213
Pythia, 50

Ra, 129, 132–33, 208–9
Ra Ho Tep (historical), 124–25,
196–97, 221–22
and channeled personality, 120–
21, 125–27
date of, 26, 28, 34
and Snefru, 32–33
Ra Ho Tep personality, 187–89,
190, 221
death of, 221
description of mushrooms, 197
first mention in Harry's drawings,
26–27
and historical person, 120–21,
125–27
intelligent direction behind,
169–72
and sacred-mushroom rites, 129–38
in trance material, 47
red crown plant, 3, 35, 159–60, 180,
183
red jars, 159

reincarnation explanation for
drawings, 162–63, 169–70
Ritual Text (from Pyramid Texts),
204–13
RNN.UT.T, 212
Round Table Foundation, 175
royal acquaintance, 125–26

sacred mushroom cult, 58–59
sacred mushroom rites, 137,
159–60, *160*
from ancient Syria, 156–59
functions of, 218
in Pyramid Texts, 195–213
and Ra Ho Tep personality,
129–38
sacred mushroom/sacred
mushrooms, 2–3, 18, 22–24,
129–38, 159–60, *160*, 195–213,
218. *See also Amanita muscaria*
cults of, 58–59
of the Druids, 74–75
use of, 131–32
salt, 132
seal with mushroom, 156–58, *156*,
195
"shade of his bush, the," 133–34
shamanism, and detachable soul
concept, 20–21
shamans, 142
call for becoming a shaman, 167
preparation for becoming, 167–68
use of Amanita muscaria, 95
SHU, 69, 133, 149, 199–201, 202
Siberian shamanism, 20–21, 167–68

INDEX **229**

"significant" statistical result, 63

slugs, and *Amanita muscaria*, 93–94

Smith Surgical Papyrus, The, 153–54

Snefru, 32–33, 145, 159, 221

soul, detachable, 20–21

souls of Nekhen, 166

spirits, 112, 171

SRK, 207

ŠSMT, 205–6

Stone, Betty, 10

Stone, Harry, 162, 178. *See also*
   hieroglyphic drawings by Harry
   Stone; trance statements by
   Harry Stone
   age-regression of, 117–20
   and *Amanita muscaria*, 99–100
   authenticity questions, 116–17,
      120–21, 126, 161–62
   biography, 41–42
   comes to Maine for experiments,
      62–69
   decides to leave experiments,
      109–10
   described, 41, 108–9
   explanations for drawings /
      statements, 27, 161–63, 169–70
   first session, 1–2, 17–20
   four stages of relationship with
      author, 107–9
   further early meetings with,
      40–49
   gold pendant blindfold test on,
      43–44
   history with ESP, 41–42
   last trance of, 138

and Maine-Mexico ESP
   experiment, 73–78
not clairvoyant, 64
reality of his trances, 49
Scrabble blocks spellings, 81–82
as sculptor, 108
spontaneous trances of, 42–43
summary of his exchanges with
   the author, 1–3
takes ESP test, 42
telepathy ability of, 64–65
Syriac medicine, and *Amanita
   muscaria*, 144–46

Taoism, 141

Tehuti, 34, 68–69, 128, 130, 188
   described, 31

TEHUTI AKH, 31

telepathic explanation for drawings,
   27, 161–62, 170

telepathy experiments, 64–66,
   175–78
   of the author, in the Army, 12

ṬHNUT, 211

TIRIAN, 78

Tiy, Queen, 10

trance cases, woman who spoke
   ancient Egyptian, 9

trance statements by Harry Stone
   Sept. 4, 1954, 45–49
   June 20, 1955, 133–34
   July 1, 1955, 135
   July 3, 1955, 135
   July 27, 1955, 136–37

230   INDEX

August 7, 1955, 137
enormous human figure, 136

TRT, 210–11
Tuat, 131

universal field of intelligence,
171–72
UPT, 154
Utterance 300 (Pyramid Texts),
195–99
Utterance 301 (Pyramid Texts),
199–204

von Strahlenberg, Johan, 2, 23

Wasson, R. Gordon, 105
his initiation into the mushroom
rite, 85
and ritual use of mushrooms,
57–59
wood, 146–47

Yayash, 147–48

Zeitgeist, 171
Zoser, King, 159–60, 220–21

About the Author

ANDRIJA PUHARICH, M.D. (February 19, 1918–January 3, 1995), was a medical and parapsychological researcher, medical inventor, physician, and author. He originally studied for a career in journalism but later decided that medicine, and particularly the workings of the human mind, was of more interest to him. Completing his studies at Northwestern University's medical school in 1947, he set up his own laboratory in Maine for the study of extrasensory perception. From 1953 to 1955, Puharich served in the Army Medical Corps at the Army Chemical Center, Edgewood Arsenal, in Maryland, and was presenting papers on the possible usefulness of paranormal phenomena in the field of intelligence work. This piqued his interest in hallucination-producing mushrooms and led to the publication of *The Sacred Mushroom* in 1959. He then went on to publish *Beyond Telepathy* in 1962.

BOOKS OF RELATED INTEREST

Microdosing with Amanita Muscaria
Creativity, Healing, and Recovery with the Sacred Mushroom
by Baba Masha, M.D.

Examining the findings from the first international study on the medicinal effects of microdosing with *Amanita muscaria*, the psychoactive fly agaric mushroom, Baba Masha, M.D., documents how more than 3,000 volunteers experienced positive outcomes for a broad range of health conditions as well as enhanced creativity and sports performance.

Sacred Mushroom of Visions: Teonanácatl
A Sourcebook on the Psilocybin Mushroom
Edited by Ralph Metzner, Ph.D.

Sacred Mushroom of Visions: Teonanácatl describes the experiences of psychoactive mushroom users and how the use of the psilocybe mushroom spread from Mexico to North America, Asia, and Europe. Firsthand accounts of the controversial Harvard Psilocybin Project, the use of psilocybin in psychotherapy, and studies on the treatment of obsessive compulsive disorder (OCD) are provided.

Magic Mushroom Explorer
Psilocybin and the Awakening Earth
by Simon G. Powell

Psilocybin is an invaluable natural resource for spiritually revivifying the human psyche and reconnecting us to the vast intelligence of Nature. This book reveals how the Earth's psychedelic medicines can bring the human race back from the brink of ecological and existential disaster.

Scan the QR code and save 25% at InnerTraditions.com. Browse over 2,000 titles on spirituality, the occult, ancient mysteries, new science, holistic health, and natural medicine.

— SINCE 1975 • ROCHESTER, VERMONT —

InnerTraditions.com • (800) 246-8648